Lecture Notes in Mathematics

A collection of informal reports and seminars
Edited by A. Dold, Heidelberg and B. Eckmann, Zürich

107

A. Peyerimhoff
University of Marburg, Marburg/Lahn

Lectures on Summability

Springer-Verlag
Berlin · Heidelberg · New York 1969

Preface

These Lecture Notes contain the material which I covered in courses on summability held at the University of Utah, Salt Lake City, in 1967 and at the University of Marburg in 1967/68. The motivation for selecting this material was two-fold: To acquaint the students with the most important parts of the theory and to lead them to a point from which they could start a master's or Ph. D. thesis in some of the fields of my own research. Functional analysis was not a prerequisite for these courses, and so I had to exclude those subjects that need detailed knowledge of this kind. I have furnished some information on the literature in the context of the courses, but I have not aimed at a complete bibliography. In fact, this is almost impossible in view of the large number of publications in summability theory during the last seventy years.

I want to express my warmest thanks to all who have helped to prepare this booklet. Part of the manuscript was read in detail by Dr. H.-H. Körle, who suggested numerous improvements. My thanks are also due to professors H.-E. Richert and D. H. Tucker as well as to cand. math. J. Brüning, Dr. H. Fiedler, Dr. W. Miesner, and Dr. R. Trautner. A first version of these notes (covering the Utah course) was very carefully typed by Mrs. Ruth Andersen, and Mrs. Helga Runckel is responsible for the excellent typing of the manuscript in its present form. Finally I owe acknowledgement to a "silent partner". Professor W. B. Jurkat, though not actually engaged in writing the manuscript, had a great influence on it through many discussions during our collaboration.

Alexander Peyerimhoff

Introductory Remarks

In 1890 E. Cesàro [14] proved the following theorem:

If $\Sigma a_n = A$ and $\Sigma b_n = B$,

and if $c_n = a_0 b_n + a_1 b_{n-1} + \ldots + a_n b_0$, $C_n = c_0 + \ldots + c_n$,

then $\frac{1}{n+1}(C_0 + C_1 + \ldots + C_n) \to AB$.

The assumptions of Cesàro's Theorem do not imply the convergence of Σc_n (counter-example $a_n = b_n = (-1)^n / \sqrt{n+1}$), they only imply the convergence of the arithmetical means of the partial sums of Σc_n .

We sketch a proof of Cesàro's Theorem.

Lemma 0.1. If $s_n \to s$, then $\sigma_n = \dfrac{s_0 + \ldots + s_n}{n+1} \to s$.

(We omit the simple proof.)

Proof of Cesàro's Theorem. A short calculation shows that

$$\frac{1}{n+1}\sum_{\nu=0}^{n} C_\nu = \frac{1}{n+1}\sum_{\nu=0}^{n} A_\nu B_{n-\nu} = \frac{B}{n+1}\sum_{\nu=0}^{n} A_\nu + \frac{1}{n+1}\sum_{\nu=0}^{n} A_\nu (B_{n-\nu} - B) = I + II$$

$(A_n = a_0 + \ldots + a_n \; , \; B_n = b_0 + \ldots + b_n)$

and it follows from Lemma 0.1. that $I \to BA$.

We have

$$|II| \leq \frac{1}{n+1}\sum_{\nu=0}^{n} |A_{n-\nu}| \, |B_\nu - B| = \frac{O(1)}{n+1}\sum_{\nu=0}^{n} |B_\nu - B| = o(1)$$

again by Lemma 0.1. .

A consequence of Cesàro's Theorem and Lemma 0.1. is: If $\Sigma a_n = A$, $\Sigma b_n = B$, and if Σc_n converges, then $\Sigma c_n = AB$ (there is either convergence of Σc_n to the "right value", or no convergence at all). This fact was discovered earlier (in 1826) by N. H. Abel [1]. He proved it in the following way: For $|x| < 1$ we have (well known properties of power series are used)

$$c(x) = \sum_0^\infty c_n x^n = \sum_0^\infty a_n x^n \sum_0^\infty b_n x^n = a(x)\, b(x) ,$$

and it follows from Abel's limit theorem, that for $x \to 1-0$,

$$\begin{array}{ccc} c(x) & = & a(x)\ b(x) \\ \downarrow & & \downarrow \quad \downarrow \\ C & = & A \quad B \end{array}$$

We mention in passing a consequence of Lemma 0.1. for geometrical means: If $0 < t_n \to t > 0$, then $\sqrt[n]{t_1 t_2 \dots t_n} \to t$ (this follows if we write $s_n = \log t_n$). For an example take $(\frac{n+1}{n})^n = t_n \to e$. Then

$$\sqrt[n]{\frac{(n+1)^n}{n!}} = \sqrt[n]{t_1 \cdots t_n} \to e$$

and it follows that $n! = (n+1)^n e^{-n} (1+o(1))^n$
(Rudimentary form of "Stirling's formula".)

We return to a discussion of Cesàro's Theorem. While ordinary convergence cannot handle the series Σc_n, another limit process, namely the convergence of the arithmetic means, yields a result. This leads to the following

<u>Definition.</u> A sequence $\{s_n\}$ is called <u>limitable</u> C_1 (i.e. in the sense of Cesàro) to the value s, we write $s_n \to s(C_1)$, if

$$\sigma_n = \frac{s_0 + \dots + s_n}{n+1} \to s \qquad (n \to \infty)$$

Lemma 0.1. states that every convergent sequence is also limitable C_1, in other words: Limitability C_1 is a generalization of the notion of convergence.

In 1904 L. Fejér [17] proved, that the arithmetical means of the partial sums of the Fourier series of a continuous function converge (whereas the Fourier series itself need not be convergent). This is another example that shows how useful the notion of limitability C_1 is, and in the following chapter we will investigate this generalization of convergence in more detail.

CHAPTER I. Cesàro means

1. The C_1 - Method

Theorem I. 1. If $s_n \to s\ (C_1)$, then $s_n = o(n)$. If $0 < \lambda_n \neq O(1)$, then there is a sequence $s_n \to s\ (C_1)$, such that $s_n \neq o(\frac{n}{\lambda_n})$ (i. e. the estimate $s_n = o(n)$ is best possible).

Proof. If $\sigma_n = \frac{s_0 + \ldots + s_n}{n+1}$ then $s_n = (n+1)\,\sigma_n - n\,\sigma_{n-1}$ $(n \geq 1)$. It follows from

$\sigma_n \to s$, that $\frac{s_n}{n} = (1+\frac{1}{n})\ \sigma_n - \sigma_{n-1} \to s - s = 0$.

Given $0 < \lambda_n \neq O(1)$, there is a sequence n_i of natural numbers such that $n_{i+1} \geq n_i + 2$ and $\lambda_{n_i} \uparrow \infty$. Now let $\sigma_{n_i} = \frac{1}{\sqrt{\lambda_{n_i}}}$, $\sigma_n = 0\ (n \neq n_i)$ It follows that the sequence s_n defined by $s_n = (n+1)\,\sigma_n - n\,\sigma_{n-1}$ is limitable C_1 to zero and that

$$\frac{\lambda_{n_i}}{n_i} s_{n_i} = \lambda_{n_i} \left(1+\frac{1}{n_i}\right) \sigma_{n_i} = \left(1+\frac{1}{n_i}\right) \sqrt{\lambda_{n_i}} \to \infty ,$$

i.e. $s_n = o(\frac{n}{\lambda_n})$ is false.

Theorem I. 1. shows that a sequence limitable C_1 has the order of growth $o(n)$. On the other hand, not every sequence $s_n = o(n)$ is limitable C_1. To verify this prove: If $s_n \uparrow \infty$ then

$$\sigma_n = \frac{s_0 + \ldots + s_n}{n+1} \uparrow \infty .$$

There are even bounded sequences which are not limitable C_1. As an example we consider the following sequence:

$$s_n \equiv \{0 \underset{1}{\overset{1}{\smile}} \ldots \underset{1}{\overset{10}{\smile}} 0\ 0 \ldots \underset{0}{\overset{100}{\smile}} 1\ 1 \ldots \underset{1}{\overset{1000}{\smile}} 0\ 0 \ldots \underset{0}{\overset{10000}{\smile}} \ldots \} .$$

Here,

$$\sigma_{10^{2k+1}} = \frac{\ldots \overset{10^{2k}}{0} + 1 + 1 + \ldots + \overset{10^{2k+1}}{1}}{10^{2k+1} + 1} \geq \frac{10^{2k+1} - 10^{2k}}{10^{2k+1} + 1} \to \frac{9}{10} \qquad (k \to \infty) ,$$

$$\sigma_{10^{2k}} = \frac{\ldots \overset{10^{2k-1}}{1} + 0 + 0 + \ldots + \overset{10^{2k}}{0}}{10^{2k} + 1} \leq \frac{10^{2k-1}}{10^{2k}+1} \to \frac{1}{10} .$$

This sequence has long intervals where s_n is constant, and it is this property which destroys its limitability - as an example of the opposite kind consider $s_n = (-1)^n$, here $\sigma_{2n} = \frac{1}{2n+1}$, $\sigma_{2n+1} = 0$, and $\sigma_n \to 0$.

These observations lead to a so-called "Tauberian Theorem" (named after the Austrian mathematician A. Tauber), i.e., to the statement "If a sequence oscillates too slowly, then it cannot be limitable C_1 unless it is convergent."

<u>Definition 1.</u> A sequence $\{s_n\}$ is called <u>slowly oscillating</u>, if $s_m - s_n \to 0$ whenever $1 \leq \frac{m}{n} \to 1$ $(m, n \to \infty)$. Using ε's and δ's this is:

To every $\varepsilon > 0$ there exists $\delta = \delta(\varepsilon)$ and $N = N(\varepsilon)$ such that $|s_m - s_n| \leq \varepsilon$ if $n \geq N(\varepsilon)$ and $n \leq m \leq (1+\delta)n$.

Definition 2. A real sequence $\{s_n\}$ is called slowly decreasing, if $\liminf s_m - s_n \geq 0$ whenever $1 \leq \frac{m}{n} \to 1$ $(m, n \to \infty)$. Using ε's and δ's this is:
To every $\varepsilon > 0$ there exists $\delta = \delta(\varepsilon)$ and $N = N(\varepsilon)$ such that $s_m - s_n \geq -\varepsilon$ if $n \geq N(\varepsilon)$ and $n \leq m \leq (1+\delta)n$.

A sequence $\{s_n\}$ is called slowly increasing, if $\{-s_n\}$ is slowly decreasing. If $\{s_n\}$ is slowly oscillating, then $\{\mathrm{Re}\, s_n\}$ and $\{\mathrm{Im}\, s_n\}$ are slowly oscillating.

Theorem I. 2. If $s_n \to s(C_1)$ and $\{s_n\}$ is slowly decreasing, then $\{s_n\}$ is convergent.

Proof. We may assume that $s = 0$ (note that $\{s_n - s\}$ is also slowly decreasing), and this implies $\limsup s_n \geq 0$, $\liminf s_n \leq 0$. If $\{s_n\}$ is not convergent, then either $\limsup s_n > 0$ or $\liminf s_n < 0$.

Let us consider the case where $\limsup s_n > 0$ first. Here, a number $\alpha > 0$ and a sequence $\{n_i\}$, $n_i \to \infty$, exists such that $s_{n_i} \geq \alpha$. If we choose $\varepsilon = \frac{\alpha}{2}$ in Definition 2, then $\delta > 0$ and N exist such that $s_m \geq s_{n_i} - \frac{\alpha}{2} \geq \frac{\alpha}{2}$ for $n_i \leq m \leq m_i = [n_i(1+\delta)]$, $n_i \geq N$, and from

$$(1) \qquad \sigma_{m_i} - \frac{n_i + 1}{m_i + 1}\sigma_{n_i} = \frac{s_{n_i+1} + \ldots + s_{m_i}}{m_i + 1} \geq \frac{\alpha}{2}\,\frac{m_i - n_i}{m_i + 1}$$

follows, by $i \to \infty$, the contradiction $0 \geq \frac{\alpha}{2}\,\frac{\delta}{1+\delta}$.

If $\liminf s_n < 0$, then we have $s_{n_i} \leq \alpha < 0$. If we choose $\varepsilon = \frac{\alpha}{2}$ in Definition 2, and if δ and M are the corresponding numbers, then it follows that $s_{n_i} - s_n \geq \frac{\alpha}{2}$ for $m_i = [\frac{n_i}{1+\delta}] \leq n \leq n_i$ (note that $n_i \leq (1+\delta)n$), and the contradiction is obtained as before (interchange n_i and m_i in (1)).

Remarks. A series $\Sigma\, a_n$ is called summable C_1 if the sequence $\{s_n\}$ $(s_n = \sum_{\nu=0}^{n} a_\nu)$ is limitable C_1.

A sequence $\{s_n\}$ is slowly oscillating when $a_n = O(\frac{1}{n})$; this follws from

$$|s_m - s_n| = |a_{n+1} + \ldots + a_m| \leq K(\frac{1}{n+1} + \ldots + \frac{1}{m}) \leq K\frac{m-n}{n} = K(\frac{m}{n} - 1) \qquad (m \geq n)$$

A similar argument shows that $\{s_n\}$ is slowly decreasing when $n a_n \geq -K$ for some $K > 0$.

In 1897 A. Tauber [101] discovered a theorem which is closely related to the following one: If Σa_n is summable C_1, and if $a_n = o(\frac{1}{n})$, then Σa_n is convergent. (Obviously this theorem is contained in Theorem I. 2.) It is for this reason that theorems of this structure (namely summability, plus some additional "Tauberian" condition on the s_n's or a_n's leads to convergence) are called "Tauberian Theorems". In 1909 G. H. Hardy [23] was able to replace the condition $a_n = o(\frac{1}{n})$ by $a_n = O(\frac{1}{n})$, and in 1910 E. Landau [61] replaced it by $n a_n \geq -K$ (this is a so-called "one sided" Tauberian theorem; theorems of this type play a role in number theory). Theorem I. 2 was proved in 1925 by R. Schmidt [93] (see also Landau [62]).

By Theorem I. 1, a sequence which is limitable C_1 cannot increase too fast. Theorem I. 2 shows that it cannot oscillate too slowly unless it is convergent. We may demonstrate this by the following diagram.

convergent sequences	sequences slowly oscillating, but not convergent	sequences limitable C_1, but not convergent	sequences $\neq o(n)$

There is a certain "zone" between the convergent sequences and the divergent and C_1-limitable sequences where no limitability takes place.

<u>Problems and examples</u>

1. Show that a series Σa_n is summable C_1 to s iff $\sigma_n = \sum_{\nu=0}^{n} (1 - \frac{\nu}{n+1}) a_\nu \to s \quad (n \to \infty)$.

2. Show that a series Σa_n is summable C_1 to s iff $\sum_{n=0}^{\infty} (\sum_{\nu=n}^{\infty} \frac{a_\nu}{\nu+1})$ exists and has the value s. (For related theorems and literature see e. g. H. C. Chow [16].)

3. If $s_n = a_0 + \dots + a_n$, then $\sigma_n = s_n - \frac{1}{n+1} \sum_{\nu=0}^{n} \nu a_\nu$. It follows that $\sigma_n \to s$ implies $s_n \to s$ if $a_n = o(\frac{1}{n})$.

4. Show that Σz^n is summable C_1 iff $|z| \leq 1$, $z \neq 1$.

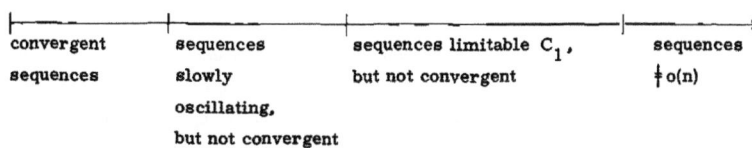

5. Show that $\Sigma\,(n+1)\,z^n$ is summable C_1 iff it is convergent.

6. Show that $\Sigma\,\sqrt{n+1}\,(-1)^n$ is summable C_1 .

7. Show that $\{\sin\log n\}$ $(n \geq 1)$ is not limitable C_1 (observe that this sequence is slowly oscillating) .

2. The C_k - Method

Given a sequence $\{s_n\}$ it may happen that it is not limitable C_1 , but the sequence $\{\sigma_n\}$ may be limitable C_1 , in other words, the iteration of the method C_1 may produce a convergent sequence. If $\{s_n\}$ is transformed into a convergent sequence after a k-fold application of the method C_1 , we call $\{s_n\}$ limitable by the method $H^k = (C_1)^k$; H refers to O. Hölder who introduced this idea in 1882 , see [37] . (See problem 4 at the end of this section.)

In some respects the method H^k is not very convenient to handle (try to express the H^k-Transformation of s_n directly in terms of s_n !). There is another generalization of the C_1 method - the Cesàro method C_k - which is easier to handle (and which is, by a theorem of K. Knopp and W. Schnee, equivalent to the method H^k ; this will be proven in section 7 of chapter II) .

In $\sigma_n = \dfrac{s_0 + \ldots + s_n}{n+1}$ the numerator is the n-th term of the summatorical sequence of $\{s_n\}$, we may write $S_n^1 = s_0 + \ldots + s_n$, while the denominator is formed from the sequence $\{1\}$ in the same way. We iterate this procedure by writing

$$\sigma_n^{(2)} = \frac{S_0^1 + \ldots + S_n^1}{1+2+\ldots+(n+1)} = \frac{S_n^2}{\binom{n+2}{n}}$$

i.e. $\sigma_n^{(2)}$ is the quotient of the n-th terms of the second summatorical sequences of $\{s_n\}$ and $\{1\}$. An iteration of this process leads to

$$\sigma_n^{(k)} = \frac{S_n^k}{\binom{n+k}{n}} \qquad (S_n^k \text{ is the n-th term of the k-fold summatorical sequence of } \{s_n\})$$

which defines Cesàro's method C_k , and $\{s_n\}$ is called limitable C_k to s if $\sigma_n^k \to s$.

In order to express s_n^k in terms of s_n we mention the following formal relation

$$\frac{1}{(1-x)} \sum_{\nu=0}^{\infty} s_\nu x^\nu = \sum_{\nu=0}^{\infty} s_\nu^1 x^\nu ,$$

and by iteration

$$\frac{1}{(1-x)^k} \sum_{\nu=0}^{\infty} s_\nu x^\nu = \sum_{\nu=0}^{\infty} s_\nu^k x^\nu .$$

But $\frac{1}{(1-x)^k} = \sum_{\nu=0}^{\infty} \binom{\nu+k-1}{\nu} x^\nu$ and this leads (multiplication of power series!) to the expression

$$s_n^k = \sum_{\nu=0}^{n} \binom{n-\nu+k-1}{n-\nu} s_\nu .$$

A short induction-type proof shows that this is actually the k-th summatorical sequence. Thus we have the following expression:

$$\sigma_n^{(k)} = \frac{1}{\binom{n+k}{n}} \sum_{\nu=0}^{n} \binom{n-\nu+k-1}{n-\nu} s_\nu ,$$

and this formula is meaningful for any (real) $k > -1$. This is the definition of Cesàro's method in its final form. Any further investigation of this method is facilitated by the Toeplitz Theorem which will be proven in the next chapter.

Problems It is customary to write $\binom{n+\alpha}{n} = A_n^\alpha$.

1. Show that $\frac{A_n^\alpha}{(n+1)^\alpha} \rightarrow \frac{1}{\Gamma(\alpha+1)}$ $\left(\frac{1}{\Gamma(-n)} = 0 \text{ for } n = 0, 1, 2, \dots\right)$.

2. Show that $\sum_{\nu=0}^{n} A_{n-\nu}^\alpha A_\nu^\beta = A_n^{\alpha+\beta+1}$.

3. Show that $s_n = \sum_{\nu=0}^{n} A_{n-\nu}^{-k-1} A_\nu^k \sigma_\nu^{(k)}$ $(k > -1)$.

4. Show that $\Sigma a_n = a(C_1)$, $\Sigma b_n = b(C_1)$ implies $\Sigma c_n = ab(C_3)$.
 (A more general result is: If Σa_n , Σb_n are summable H^p and H^q , respectively, then the product-series Σc_n is summable H^{p+q+1} .)

5. Show that $\Sigma (n+1)(-1)^n$ is summable C_2 .

6. Show that $\frac{\sigma_0+\dots+\sigma_n}{n+1} = \frac{1}{n+1}\sum_{\nu=0}^{n}(h_n - h_{\nu-1})s_\nu$ $\quad (h_n = 1 + \frac{1}{2} + \dots + \frac{1}{n},\ h_{-1} = 0)$
(H^2 method).

CHAPTER II Matrix transformations

1. Theorems of Toeplitz and Schur

Lemma II. 1. Let $(a_{n\nu})$ $(n, \nu = 0, 1, \dots)$ be a matrix with the properties

(1) $$\sum_{\nu=0}^{\infty} |a_{n\nu}| < \infty \qquad (n = 0, 1, \dots)$$

(2) $$\sum_{\nu=0}^{k} |a_{n\nu}| = o\left(\sum_{\nu=k+1}^{\infty} |a_{n\nu}|\right) \qquad (n \to \infty, \text{ for every fixed } k).$$

Then, there are sequences $0 < \nu_1 < \nu_2 < \dots,$ $0 < n_1 < n_2 < \dots$ such that

(3) $$\sum_{\nu=\nu_i}^{\nu_{i+1}-1} |a_{n_i,\nu}| \geq \frac{i-1}{i}\sum_{\nu=0}^{\infty} |a_{n_i,\nu}| \qquad (i = 1, 2, 3 \dots).$$

Proof. We proceed by induction. Inequality (3) is true for $i=1$ and arbitrarily chosen numbers ν_1, ν_2, n_1. Given ν_i, ν_{i+1} and n_i we choose n_{i+1}, ν_{i+2} such that

$$\sum_{\nu_{i+1}}^{\infty} |a_{n_{i+1},\nu}| \geq 2(i+1) \sum_{0}^{\nu_{i+1}-1} |a_{n_{i+1},\nu}|$$

and

$$\sum_{\nu_{i+1}}^{\infty} |a_{n_{i+1},\nu}| \geq 2(i+1) \sum_{\nu_{i+2}}^{\infty} |a_{n_{i+1},\nu}|$$

(this is possible because of (1) and (2)). It follows that

$$\sum_{\nu_{i+1}}^{\nu_{i+2}-1} |a_{n_{i+1},\nu}| = \frac{i}{i+1} \sum_{\nu_{i+1}}^{\nu_{i+2}-1} |a_{n_{i+1},\nu}| + \frac{1}{2(i+1)} \sum_{\nu_{i+1}}^{\infty} |a_{n_{i+1},\nu}|$$

$$+ \frac{1}{2(i+1)} \sum_{\nu_{i+1}}^{\infty} |a_{n_{i+1},\nu}| - \frac{1}{i+1} \sum_{\nu_{i+2}}^{\infty} |a_{n_{i+1},\nu}|$$

$$\geq \frac{i}{i+1} \sum_{\nu_{i+1}}^{\nu_{i+2}-1} |a_{n_{i+1},\nu}| + \sum_{0}^{\nu_{i+1}-1} |a_{n_{i+1},\nu}| + \left(1-\frac{1}{i+1}\right) \sum_{\nu_{i+2}}^{\infty} |a_{n_{i+1},\nu}|$$

$$\geq \frac{i}{i+1} \sum_{0}^{\infty} |a_{n_{i+1},\nu}| \quad .$$

Theorem II. 1. (O. Toeplitz, see [102], [58], [97]) Given a matrix $(a_{n\nu})$ $(n, \nu = 0, 1, \ldots)$ we introduce the conditions

(RN) $\quad \sum_{\nu=0}^{\infty} |a_{n\nu}| = O(1)$ $\quad$ ($n \to \infty$; Row-Norm condition) ,

(RS_α) $\quad \sum_{\nu=0}^{\infty} a_{n\nu} \to \alpha$ $\quad$ ($n \to \infty$; Row-Sum condition) ,

(C_{α_ν}) $\quad a_{n\nu} \to \alpha_\nu$ $\quad$ ($n \to \infty$, ν fixed; Column condition) .

Writing $\sigma_n = \sum_{\nu=0}^{\infty} a_{n\nu} s_\nu$, we have the following statements:

(1^0) $s_n = O(1)$ implies $\sigma_n = O(1)$ iff (RN) holds (the statement $\sigma_n = O(1)$ presupposes that σ_n exists for $n = 0, 1, \ldots$; a similar remark holds in all following cases) ,

(2°) $s_n = o(1)$ implies $\sigma_n = O(1)$ iff (RN) holds,

(3°). $\{s_n\}$ convergent implies $\{\sigma_n\}$ convergent iff (RN), (RS_α) and (C_{α_ν}) holds;

in this case $\lim \sigma_n = (\alpha - \Sigma\alpha_\nu) \lim s_n + \sum_{\nu=0}^{\infty} \alpha_\nu s_\nu$, and, therefore,

$\lim \sigma_n = \lim s_n$ iff $\alpha = 1$, $\alpha_n = 0$,

(4°) $s_n = o(1)$ implies $\{\sigma_n\}$ convergent iff (RN) and (C_{α_ν}) holds.

A matrix with the property (3°) is called <u>convergence-preserving</u>, and is called <u>regular</u> if, in addition, $\lim \sigma_n = \lim s_n$.

<u>Proof.</u> The sufficient part of each statement is easily verified. (Note that (RN) and (C_{α_ν}) imply $\Sigma |\alpha_\nu| < \infty$.)

We turn to the necessity of (RN) in (2°). First we have $\sum_{\nu=0}^{\infty} |a_{n\nu}| < \infty$ for every n since $\{\sigma_n\}$ exists. (This follows from the Abel-Dini Theorem: If $d_n \geq 0$ and $\sum_0^\infty d_n = \infty$, then $\sum_0^\infty \frac{d_n}{D_n} = \infty$, $D_n = d_0 + \ldots + d_n + 1$.) Furthermore, the sequence $s^{(p)} = (0\ 0 \ldots \overset{p}{1}\ 0 \ldots)$ is transformed into $\sigma_n = a_{np}$, and therefore $a_{np} = O(1)$ $(n \to \infty,\ p$ fixed). We assume that (RN) is not true, and we may assume (by omitting rows in the matrix) that $n^2 \leq \sum_{\nu=0}^{\infty} |a_{n\nu}|$.

It follows from Lemma II. 1. that $\sum_{\nu_i}^{\nu_{i+1}-1} |a_{n_i,\nu}| \geq \frac{i-1}{i} \sum_0^\infty |a_{n_i,\nu}|$. Let $s_\nu = \frac{2}{i} \operatorname{sign} a_{n_i,\nu}$ for $\nu_i \leq \nu < \nu_{i+1}$ $(i = 2, 3, \ldots)$, $s_\nu = 0$ otherwise. We have $s_n \to 0$ and $|s_n| \leq 1$. But

$$\Big|\sum_0^\infty a_{n_i,\nu} s_\nu\Big| \geq \frac{2}{i} \sum_{\nu=\nu_i}^{\nu_{i+1}-1} |a_{n_i,\nu}| - \sum_0^{\nu_i - 1} |a_{n_i,\nu}| - \sum_{\nu_{i+1}}^{\infty} |a_{n_i,\nu}|$$

$$= \frac{2+i}{i} \sum_{\nu=\nu_i}^{\nu_{i+1}-1} |a_{n_i,\nu}| - \sum_0^\infty |a_{n_i,\nu}| \geq \Big(\frac{(i+2)(i-1)}{i^2} - 1\Big) \sum_0^\infty |a_{n_i,\nu}|$$

$$\geq \Big(\frac{1}{i} - \frac{2}{i^2}\Big)\ i^2 \to \infty ,$$

and this shows that (RN) is necessary. It follows from this result that (RN) is also necessary in (1°), (3°) and (4°). The remaining conditions follow immediately from the transformations of the special sequences $s^{(p)}$ and $(1, 1, 1, \ldots)$.

Theorem II. 2 (Schur, 1921 see [97]). $s_n = O(1)$ implies the convergence of $\sigma_n = \sum_{\nu=0}^{\infty} a_{n\nu} s_\nu$ iff (RN), (C_{α_ν}) and $\sum_{\nu=0}^{\infty} |a_{n\nu} - \alpha_\nu| \to 0 \quad (n \to \infty)$ holds. The matrix is called convergence generating in this case.

Proof. The sufficiency is obvious. It follows from (4^0) in Theorem II. 1. that (RN) and (C_{α_ν}) are necessary.

Let $a^*_{n\nu} = a_{n\nu} - \alpha_\nu$; this is again a matrix which transforms $s_n = O(1)$ into a convergent sequence.

Assume that $\sum_{\nu=0}^{\infty} |a^*_{n\nu}| \to 0 \quad (n \to \infty)$ is not true. By omitting rows of the matrix we may assume that $\sum_{\nu=0}^{\infty} |a^*_{n\nu}| \geq \delta$ for some $\delta > 0$. If $s_\nu = (-1)^i \operatorname{sign} a^*_{n_i \nu}$ ($\nu_i \leq \nu < \nu_{i+1}$, $s_\nu = 0$ otherwise) where ν_i, n_i are given by Lemma II. 1. (note that $a^*_{n\nu} \to 0$, $n \to \infty$, ν fixed), it follows that

$$(-1)^i \sum_{\nu=0}^{\infty} a^*_{n_i \nu} s_\nu \geq \sum_{\nu=\nu_i}^{\nu_{i+1}-1} |a^*_{n_i,\nu}| - \sum_{0}^{\nu_i - 1} |a^*_{n_i \nu}| - \sum_{\nu_{i+1}}^{\infty} |a^*_{n_i,\nu}|$$

$$= 2 \sum_{\nu=\nu_i}^{\nu_{i+1}-1} |a^*_{n_i,\nu}| - \sum_{0}^{\infty} |a^*_{n_i,\nu}| \geq \left(2\,\frac{i-1}{i} - 1\right) \sum_{0}^{\infty} |a^*_{n_i,\nu}| \geq \delta\left(1 - \frac{2}{i}\right) ,$$

and this implies that $\sum_{0}^{\infty} a^*_{n_i,\nu} s_\nu$ is not convergent.

(Incidentally, Schur's theorem is equivalent with the functional-analytic statement that weak and strong convergence in the space l_1 are equivalent; see for instance [5] p. 137.)

Problem. Prove the following theorem of K. Knopp and G. G. Lorentz [57] :
A matrix $A = (a_{n\nu})$ $(n, \nu = 0, 1, \ldots)$ has the property that

$$\sum_n \Big| \sum_\nu a_{n\nu} a_\nu \Big| < \infty \quad \text{whenever} \quad \sum |a_\nu| < \infty \quad \text{iff} \quad \sum_n |a_{n\nu}| = O(1) \qquad (\nu \to \infty) .$$

2. Applications of the theorems of Toeplitz and Schur

1. A regular matrix $A = (a_{n\nu})$ is not convergence-generating. In this case $\alpha_\nu = 0$, and it would follow from Schur's theorem that (for $n \to \infty$)

$$\left|\sum_{\nu=0}^{\infty} a_{n\nu}\right| \le \sum_{\nu=0}^{\infty} |a_{n\nu}| \to 0$$

but $\sum_{\nu=0}^{\infty} a_{n\nu} \to 1$ because of the regularity of A.

2. If $A = (a_{n\nu})$ is triangular (i.e. $a_{n\nu} = 0$ for $\nu > n$), and if $a_{nn} \neq 0$ (in what follows, a triangular matrix with non-vanishing diagonal terms will be called normal), then the transformation $\sigma_n = \sum_{\nu=0}^{n} a_{n\nu} s_\nu$ can be inverted; we write $s_n = \sum_{\nu=0}^{n} a'_{n\nu}\sigma_\nu$. ($A' = A^{-1} = (a'_{n\nu})$ denotes the inverse matrix of $(a_{n\nu})$.) Later we will frequently make use of the fact that $a_{nn}a'_{nn} = 1$.

Theorem II. 3. Let $(a_{n\nu})$ be normal, $\sigma_n = \sum_{\nu=0}^{n} a_{n\nu}s_\nu$ and $G_n = \sum_{\nu=0}^{n} |a'_{n\nu}|$. Then

(i) $\sigma_n = O(1)$ implies $s_n = O(G_n)$, but not $s_n = O(\frac{G_n}{\lambda_n})$ for any $\lambda_n \neq O(1)$,

(ii) $\sigma_n = o(1)$ implies $s_n = o(G_n)$ iff $a'_{n\nu} = o(G_n)$ ($n \to \infty$, ν fixed); G_n may not be replaced by $\frac{G_n}{\lambda_n}$ for any $\lambda_n \neq O(1)$ in $s_n = o(G_n)$.

Proof. (i) It follows from the Toeplitz theorem (part 1°) that

$$\frac{s_n}{G_n} = \sum_{\nu=0}^{n} \frac{a'_{n\nu}}{G_n}\sigma_\nu = O(1),$$

while

$$\lambda_n \frac{s_n}{G_n} = O(1) \qquad \text{is not}$$

always true because of $\lambda_n \frac{1}{G_n}\sum_{\nu=0}^{n} |a'_{n\nu}| = \lambda_n \neq O(1)$.

(ii) If $\sigma_n = o(1)$, then it follows from part 4° of the Toeplitz theorem that $\frac{s_n}{G_n} = o(1)$ iff $a'_{n\nu} = o(G_n)$, and the last statement of the theorem follows as in the proof of (i).

Remark. If A is regular, then, under the assumptions of (ii), $\sigma_n \to s$ implies $s_n = s + o(G_n)$ (note that $\sum_{\nu=0}^{n} a_{n\nu}(s_\nu - s) \to 0$). If $G_n \to \infty$, then the conclusion reduces to $s_n = o(G_n)$.

3. We will use the following notation. Given a matrix $A = (a_{n\nu})$ $(n, \nu = 0, 1, \ldots)$, a sequence $\{s_\nu\}$ is called A-limitable to s - we write $s_n \to s$ (A) - if $\sigma_n = \sum_{\nu=0}^{\infty} a_{n\nu} s_\nu$ exists for $n = 0, 1, \ldots$, and if $\sigma_n \to s$. A series will be called A-summable to s if the sequence of its partial sums is A-limitable to s - we write $\Sigma a_n = s$ (A) .

Given two methods $A = (a_{n\nu})$, $B = (b_{n\nu})$, one might ask for conditions which guarantee that a sequence is B-limitable whenever it is A-limitable. We write $A \subseteqq B$ for this inclusion relation Let $A = (a_{n\nu})$ be triangular and $a_{nn} \neq 0$. Then $s_n = \sum_{\nu=0}^{n} a'_{n\nu} \sigma_\nu$, where $\sigma_n \to s$ (i.e. $s_n \to s$ (A)) is limitable by the method B iff

$$\tau_n = \sum_{\nu=0}^{n} b_{n\nu} \sum_{\mu=0}^{\nu} a'_{\nu\mu} \sigma_\mu = \sum_{\mu=0}^{n} \sigma_\mu \sum_{\nu=\mu}^{n} b_{n\nu} a'_{\nu\mu}$$

is convergence preserving. With the notation $C = BA^{-1} = (\sum_{\nu=\mu}^{n} b_{n\nu} a'_{\nu\mu})$ we have

Theorem II. 4. Let A be normal, and let B be triangular. Then $A \subseteqq B$ iff $C = BA^{-1}$ is convergence preserving.

It should be noted that $A \subseteqq B$ is consistent with $s_n \to s$ (A) , $s_n \to s'$ (B) , $s \neq s'$. Equality $s = s'$ holds if C is regular.

If $A \subseteqq B$ and $B \subseteqq A$, then we write $A \approx B$ and call A and B equivalent .

Examples. The methods C_α $(\alpha \geq 0)$ are regular, and we have $(C'_\alpha)_{n\nu} = A_{n-\nu}^{-\alpha-1} A_\nu^\alpha$. There are two positive constants ε , K such that $\varepsilon \leq \frac{G_n}{(n+1)^\alpha} \leq K$. Furthermore, $|A_{n-\nu}^{-\alpha-1} A_\nu^\alpha| = o((n+1)^\alpha)$, and it follows that $s_n \to 0$ (C_α) implies $s_n = o((n+1)^\alpha)$, but not $s_n = o(\frac{(n+1)^\alpha}{\lambda_n})$ for any $\lambda_n \neq O(1)$ (as for the case $\alpha = 1$ see Theorem I. 1., page 4). If $0 \leq \alpha < \beta$, then $C_\beta \supseteqq C_\alpha$ (and we have $s' = s$) . This follows from

$$(C_\beta C'_\alpha)_{n\nu} = \sum_{\mu=\nu}^{n} A_{n-\mu}^{\beta-1} A_{\mu-\nu}^{-\alpha-1} \frac{A_\nu^\alpha}{A_n^\beta} = \frac{A_{n-\nu}^{\beta-\alpha-1} A_\nu^\alpha}{A_n^\beta} \quad .$$

Problems.

1. Show that to every regular matrix a sequence consisting only of 0's and 1's exists which is not limitable by this matrix.

2. Show that $C_1 \subseteqq A$ (A regular and triangular) holds iff $\sum_{\nu=0}^{n} (\nu+1) |a_{n\nu}-a_{n,\nu+1}| = O(1)$ $(n \to \infty)$.

3. Special matrix methods

In this section we will discuss some of the most important special methods. A matrix $A = (a_{n\nu})$ $(n, \nu = 0, 1, \dots)$ generates a matrix method of summation. Thus the matrix $(A_{n-\nu}^{\alpha-1} / A_n^{\alpha})$ generates the Cesàro method C_α. Sometimes we will use the same symbol for a matrix and for the method generated by it. For instance, we will write $(C_\alpha)_{n\nu} = A_{n-\nu}^{\alpha-1} / A_n^{\alpha}$.

1. (Weighted) arithmetical means are of the form

$$\sigma_n = \frac{1}{P_n} \sum_{\nu=0}^{n} p_\nu s_\nu \qquad (P_n = p_0 + \dots + p_n \neq 0) .$$

The matrix method given by $a_{n\nu} = \frac{p_\nu}{P_n}$ $(\nu \leq n)$, $a_{n\nu} = 0$ $(\nu > n)$ is denoted by M_p.

This method is regular iff $\sum_{\nu=0}^{n} |p_\nu| = O(P_n)$, $P_n \to \infty$. In particular, when $p_n > 0$, then M_p is regular iff $P_n \to \infty$. Writing $\sigma_n = \frac{1}{P_n} \sum_{\nu=0}^{n} p_\nu s_\nu$ we have $s_n = \frac{P_n}{p_n} \sigma_n - \frac{P_{n-1}}{p_n} \sigma_{n-1}$ $(\sigma_{-1} = 0$, $p_n \neq 0)$. If $p_n > 0$, then $G_n \sim \frac{P_n}{p_n}$ $(a_n \sim b_n$ means that positive constants ε, K exist such that $\varepsilon \leq \frac{b_n}{a_n} \leq K$ for n sufficiently large). It follows from Theorem II. 3. that $s_n = o(\frac{P_n}{p_n})$ is a consequence of $s_n \to 0$ (M_p), and this estimate of s_n cannot be improved.

If $\frac{P_n}{p_n} = O(1)$ (as for instance in the case $p_n = 2^n$), then $s_n \to 0$ (M_p) iff $s_n \to 0$. In this case M_p is equivalent to convergence (special case of a so-called "high indices theorem").

2. <u>Nörlund means</u> are of the form

$$\sigma_n = \frac{p_n s_0 + p_{n-1} s_1 + \dots + p_0 s_n}{P_n} \qquad (P_n = p_0 + \dots + p_n \neq 0)$$

The method thus defined is denoted by N_p (sometimes called Nörlund mean N_p), and it is regular iff $\sum_{\nu=0}^{n} |p_\nu| = O(P_n)$ and $p_n = o(P_n)$. In order to see this note that $\frac{P_{n-1}}{P_n} = \frac{P_{n-1}}{P_{n-1}} \frac{P_n - p_n}{P_n} \to 0$, and that we have $\frac{p_{n-k}}{P_n} \to 0$ by repetition. For $p_n \geq 0$ the mean N_p is regular iff $p_n = o(P_n)$.

In order to obtain the inverse of the N_p transform we write formally $p(z) = \Sigma\, p_n z^n$, $k(z) = \frac{1}{p(z)} = \Sigma\, k_n z^n$. The inverse of N_p is then given by $(N'_p)_{n\nu} = k_{n-\nu} P_\nu$ $(\nu \leq n)$. This follows from

$$\sum_{\nu=0}^{n} k_{n-\nu} P_\nu \sigma_\nu = \sum_{\nu=0}^{n} k_{n-\nu} \sum_{\mu=0}^{\nu} p_{\nu-\mu} s_\mu = \sum_{\mu=0}^{n} s_\mu \sum_{\nu=\mu}^{n} k_{n-\nu} p_{\nu-\mu} = s_n .$$

The method N_p, $p_n = A_n^{\alpha-1}$ $(\alpha > -1)$, is the Cesàro method C_α. In this case $p(z) = \frac{1}{(1-z)^\alpha}$, $k(z) = (1-z)^\alpha = \Sigma\, A_\nu^{-\alpha-1} z^\nu$ and $(C_\alpha^{-1})_{n\nu} = A_{n-\nu}^{-\alpha-1} A_\nu^\alpha$, a formula which we mentioned earlier.

For another example we consider the Nörlund mean N_p where $p_0 = 1$, $p_1 = -2$, $p_2 = p_3 = \dots = 0$. In this case $\sigma_0 = s_0$, $\sigma_n = 2s_{n-1} - s_n$, which represents a regular method. We have $p(z) = 1-2z$, $k(z) = \frac{1}{1-2z} = \Sigma\, 2^n z^n$, $k_n = 2^n$. The inverse transformation is given by

$$s_n = 2^n \sigma_0 - \sum_{\nu=1}^{n} 2^{n-\nu} \sigma_\nu \qquad (n \geq 1) \quad ;$$

if $\sigma_n \to 0$ then

$$s_n - 2^n(\sigma_0 - \sum_{\nu=1}^{\infty} \frac{\sigma_\nu}{2^\nu}) = \sum_{\nu=n+1}^{\infty} 2^{n-\nu} \sigma_\nu ,$$

and an application of the Toeplitz theorem shows that $\sum_{\nu=n+1}^{\infty} 2^{n-\nu}\sigma_\nu \to 0$. It follows that a sequence $\{s_n\}$ is limitable to zero by this method iff s_n is of the form $c\,2^n + c_n$, where c is a constant and $c_n \to 0$. In other words, this method is able to handle - essentially - only one divergent sequence, namely $\{2^n\}$. This may be considered "pathological" , and it is the purpose of the notion of "perfecticity" (which will be introduced later) to single out pathological methods of this kind. (G. H. Hardy [24] was the first to observe that methods with this pathological property exist; for further results see e. g. [80], [112] .)

Two methods A and B are called consistent if $s_n \to s$ (A) , $s_n \to s'$ (B) implies $s = s'$. We wish to show that all regular and real (p_n real) Nörlund methods are consistent. This is a consequence of the following theorem, which is of interest in itself.

Theorem II. 5 (see [114] , [46]). Let N_p be a regular and real Nörlund method, and assume that $s_n \to s$ (N_p) . Then $s(z) = \sum s_n z^n$ has a positive radius of convergence. The function s(z) permits analytic continuation onto $|z| < 1$ with the exception of poles. There exists an $\varepsilon > 0$ such that s(z) is regular for $1-\varepsilon < z < 1$ and $\lim_{x \to 1-0} (1-x)\,s(x) = s$ $(1-\varepsilon < x < 1)$.

Proof. Let $s_n \to s$ (N_p) , then $s_n = \sum_{\nu=0}^{n} k_{n-\nu} P_\nu \sigma_\nu$, $\sigma_n \to s$, and it follows for small $|z|$ that $\sum s_n z^n = k(z) \sum_{\nu=0}^{\infty} P_\nu \sigma_\nu z^\nu = \frac{1}{p(z)} \sum_{\nu=0}^{\infty} P_\nu \sigma_\nu z^\nu$ (note that $\sum p_n z^n$, $\sum P_n \sigma_n z^n$ are convergent for $|z|<1$ because of $\frac{P_{n+1}}{P_n} \to 1$, and that $p(0) = p_0 \neq 0$). This formula shows that s(z) is regular for $|z|<1$ with the exception of poles, and s(z) is regular for $z = 0$.

It follows from

$$\sum_{\nu=0}^{n} |p_\nu| = O(P_n) , \quad \frac{P_{n+1}}{P_n} \to 1 ,$$

that either $P_n \geq \delta > 0$ or $P_n \leq -\delta < 0$ (P_n is real!) for large n , $n \geq N$ say, and that $P(z) = \frac{1}{1-z}\,p(z) = \sum P_n z^n \neq 0$ for $1-\varepsilon < z < 1$. The statement $\lim_{1-\varepsilon < x \to 1-0} (1-x)s(x) = s$ follows from the Toeplitz theorem since, for $x \to 1-0$,

$$\frac{\sum_{0}^{\infty} |P_\nu|\, x^\nu}{|P(x)|} \leq \frac{|P(x)| + 2\,|\sum_{\nu=0}^{N} P_\nu x^\nu|}{|P(x)|} = 1 + o(1)$$

(observe that $|P(x)| \to \infty$) and $P_n x^n = o(\sum_0^\infty P_\nu x^\nu)$.

Let N_p, N_q be regular and real Nörlund methods and assume that $s_n \to s$ (N_p) , $s_n \to s'$ (N_q) . Then, $s = \lim (1-x)s(x) = s'$ by Theorem II. 5. . We thus have proven

Theorem II. 6. All regular and real Nörlund methods are consistent.

We conclude our discussion of Nörlund methods with a theorem on equivalence that will be used in chapter IV.

Theorem II. 7. ([92]). Let N_p, N_q be regular. Then $N_p \approx N_q$ iff

$$\frac{p(z)}{q(z)} = \sum a_n z^n \quad , \quad \frac{q(z)}{p(z)} = \sum b_n z^n \quad \text{satisfy} \quad \sum |a_n| < \infty \quad , \quad \sum |b_n| < \infty .$$

Proof. Necessity.

If $N_p \subseteq N_q$ then

$$(N_q N_p^{-1})_{n\nu} = \frac{P_\nu}{Q_n} \sum_{\rho=\nu}^{n} q_{n-\rho} k_{\rho-\nu} = \frac{P_\nu}{Q_n} b_{n-\nu}$$

satisfies (RN) (Theorems II. 4. and II. 1.), i. e. we have $\sum_{\nu=0}^{n} |P_\nu b_{n-\nu}| = O(Q_n)$, and this implies (for $\nu = n$) that $P_n = O(Q_n)$. Likewise it follows from $N_q \subseteq N_p$ that $Q_n = O(P_n)$. Therefore, if $N_p \approx N_q$, then

$$\sum_{\nu=0}^{n_0} |b_\nu| \left|\frac{P_{n-\nu}}{P_n}\right| = O\left(\frac{Q_n}{P_n}\right) \leq K \qquad (0 \leq n_0 \leq n \text{ , } K \text{ independent of } n_0 \text{ and } n).$$ For $n \to \infty$, this implies $\sum_0^{n_0} |b_\nu| \leq K$ and therefore, $\sum |b_\nu| < \infty$. It follows the same way that $\sum |a_\nu| < \infty$.

Sufficiency.

We show first that $N_q N_p^{-1}$ is regular. It follows from $\sum |a_\nu| < \infty$ that

$$|P_n| = \left|\sum_{\nu=0}^{n} a_{n-\nu} Q_\nu\right| = \left|\sum_{\nu=0}^{n} A_{n-\nu} q_\nu\right| = O(Q_n) \qquad (A_n = a_0 + \dots + a_n) ,$$

and the condition (RN) is satisfied for $N_q N_p^{-1}$ because of

$$\sum_{\nu=0}^{n} |P_\nu| |b_{n-\nu}| \leq \sum_{\nu=0}^{n} |b_{n-\nu}| \sum_{\mu=0}^{\nu} |p_\mu| = \sum_{\mu=0}^{n} |p_\mu| \sum_{\nu=\mu}^{n} |b_{n-\nu}| = O(P_n) = O(Q_n) .$$

Condition (RS_1) is obviously satisfied for $N_q N_p^{-1}$, and (C_o) follows from $b_n = o(Q_n)$ (observe that $b_n = o(1)$, whereas $\varepsilon > 0$ exists with $|Q_n| \geq \varepsilon$).

A similar proof shows that $N_p N_q^{-1}$ is regular.

3. The Euler means of a sequence $\{s_n\}$ are of the form

$$\sigma_n = \frac{1}{(p+1)^n} \sum_{\nu=0}^{n} \binom{n}{\nu} p^{n-\nu} s_\nu \qquad (p > 0) \ .$$

This method is denoted by E_p. A short calculation shows that

$$(E_p^{-1})_{n\nu} = \binom{n}{\nu} (-1)^{n-\nu} p^{n-\nu} (p+1)^\nu \ .$$

It follows that G_n (see page 14) is $(2p+1)^n$, and from $(E_p^{-1})_{n\nu} = o((2p+1)^n)$ $(n \to \infty, \nu$ fixed) and Theorem II. 3. we obtain that $s_n \to 0$ (E_p) implies $s_n = o((2p+1)^n)$. This estimate is best possible. (A first systematic treatment of E_p was given by K. Knopp [52] .)

4. Hausdorff means (introduced in [32] ; see also [38]). We begin with the problem to determine all triangular matrices A with the property $AC_1 = C_1 A$ (with the Cesàro method C_1). This identity reads

(4) $$\sum_{\mu=\nu}^{n} \frac{a_{n\mu}}{\mu+1} = \sum_{\mu=\nu}^{n} \frac{a_{\mu\nu}}{n+1} \qquad \text{for} \quad 0 \leq \nu \leq n = 0,1,2 \ldots \ .$$

It follows from (4) that

(5) $$\sum_{\mu=\nu+1}^{n} \frac{n+1}{\mu+1} a_{n\mu} = \sum_{\mu=\nu+1}^{n} a_{\mu,\nu+1} \ ;$$

(4) and (5) imply

(6) $$\left(\frac{n+1}{\nu+1} - 1\right) a_{n\nu} = \sum_{\mu=\nu}^{n-1} a_{\mu\nu} - \sum_{\mu=\nu+1}^{n} a_{\mu,\nu+1} \ .$$

It follows from formula (6) that the elements $a_{n\nu}$ $(\nu < n)$ are uniquely determined if the elements a_{nn} are given. We will next establish a formula which expresses $a_{n\nu}$ $(\nu < n)$ in terms of a_{nn} explicitely.

Using the notation

$$\Delta^0 x_n = x_n \ , \quad \Delta^1 x_n = x_n - x_{n+1} \ , \quad \Delta^{k+1} x_n = \Delta^1(\Delta^k x_n) \ , \quad \Delta^k x_n = \sum_{\nu=0}^{k} \binom{k}{\nu} (-1)^\nu x_{n+\nu}$$

we wish to show by induction (with respect to $n-\nu$) that $a_{n\nu} = \binom{n}{\nu} \Delta^{n-\nu} a_{\nu\nu}$. For $n-\nu = 0$ this formula is true. Assuming that it holds for a_{ik} , whenever $i-k < n-\nu$, we have ($a_\rho = a_{\rho\rho}$)

$$\sum_{\mu=\nu}^{n-1} a_{\mu\nu} = \sum_{\mu=\nu}^{n-1} \binom{\mu}{\nu} \Delta^{\mu-\nu} a_\nu = \sum_{\mu=\nu}^{n-1} \binom{\mu}{\nu} \sum_{\rho=\nu}^{\mu} \binom{\mu-\nu}{\rho-\nu} (-1)^{\rho-\nu} a_\rho$$

$$= \sum_{\rho=\nu}^{n-1} (-1)^{\rho-\nu} a_\rho \frac{\rho!}{\nu!(\rho-\nu)!} \sum_{\mu=\rho}^{n-1} \frac{\mu!}{(\mu-\rho)!\rho!} \ .$$

But

$$\sum_{\mu=\rho}^{n-1} \binom{\mu}{\rho} = \sum_{\mu=0}^{n-1-\rho} \binom{\mu+\rho}{\mu} = \sum_{\mu=0}^{n-1-\rho} A_\mu^\rho = A_{n-1-\rho}^{\rho+1} = \binom{n}{\rho+1} \ ,$$

and it follows that

$$\sum_{\mu=\nu}^{n-1} a_{\mu\nu} - \sum_{\mu=\nu+1}^{n} a_{\mu,\nu+1} = \sum_{\rho=\nu}^{n} a_\rho (-1)^{\rho-\nu} \left\{ \binom{\rho}{\nu}\binom{n}{\rho+1} + \binom{\rho}{\nu+1}\binom{n+1}{\rho+1} \right\}$$

$$= \sum_{\rho=\nu}^{n} a_\rho (-1)^{\rho-\nu} \binom{n}{\nu+1} \binom{n-\nu}{\rho-\nu} \ .$$

This identity and (6) show that $a_{n\nu} = \binom{n}{\nu} \Delta^{n-\nu} a_{\nu\nu}$. On the other hand, it is obvious from the foregoing proof that, for any sequence $\{a_{nn}\}$, the matrix $A = (\binom{n}{\nu} \Delta^{n-\nu} a_{\nu\nu})$ satisfies the relation $AC_1 = C_1A$.

A triangular matrix A is called a Hausdorff matrix (defining a <u>Hausdorff transformation</u>) when $AC_1 = C_1A$, i.e., iff $a_{n\nu} = \binom{n}{\nu} \Delta^{n-\nu} d_\nu$ for some sequence $\{d_\nu\}$.

We mention another representation of a Hausdorff matrix A . If $\Delta = ((-1)^\nu \binom{n}{\nu})$, $D = (d_n \delta_{n\nu})$ (i.e., D is a diagonal-matrix with the diagonal elements d_n), then

$$(\Delta D \Delta)_{n\nu} = \sum_{\nu \le \rho \le \mu \le n} \binom{n}{\mu} (-1)^\mu d_\mu \delta_{\mu\rho} \binom{\rho}{\nu} (-1)^\nu = \sum_{\mu=\nu}^{n} \frac{n!}{\mu!(n-\mu)!} \frac{\mu!}{\nu!(\mu-\nu)!} (-1)^{\mu-\nu} d_\mu$$

$$= \binom{n}{\nu} \Delta^{n-\nu} d_\nu \ ,$$

i.e., every Hausdorff matrix can be written in the form $(\Delta D \Delta)_{n\nu}$. (It follows, in particular, that $\Delta^2 = I$ with the idendity matrix I; therefore, $\Delta^{-1} = \Delta$). This representation shows that any two Hausdorff matrices commute (note that $H_1 = \Delta D_1 \Delta$, $H_2 = \Delta D_2 \Delta$ implies $H_1 H_2 = \Delta D_1 \Delta\Delta D_2 \Delta = \Delta D_1 D_2 \Delta = \Delta D_2 D_1 \Delta = H_2 H_1$. As a consequence we see that all regular Hausdorff methods are consistent.

Next we discuss some special Hausdorff methods.

a) If $d_n = \frac{1}{\binom{n+\alpha}{n}} = \alpha \frac{\Gamma(n+1)\,\Gamma(\alpha)}{\Gamma(n+\alpha+1)} = \alpha \int_0^1 t^n (1-t)^{\alpha-1} dt \qquad (\alpha > 0)$, then

$$\binom{n}{\nu} \Delta^{n-\nu} d_\nu = \binom{n}{\nu} \alpha \int_0^1 t^\nu (1-t)^{n-\nu+\alpha-1} dt = \frac{A_{n-\nu}^{\alpha-1}}{A_n^\alpha} .$$

This sequence generates the Hausdorff method C_α.

b) It follows from the foregoing example that C_1 is generated by the sequence $d_n = \frac{1}{n+1}$. The Hölder methods H^k $(k=1,2,\ldots)$, therefore, are generated by the sequence $\{(n+1)^{-k}\}$, and we have

$$(7) \qquad H^k = \Delta \left(\frac{\delta_{n\nu}}{(n+1)^k} \right) \Delta .$$

The right hand side of (7) is also meaningful when k is not an integer; (7) is Hausdorff's definition of Hölder means of arbitrary order. It follows from

$$\frac{1}{(n+1)^k} = \frac{1}{\Gamma(k)} \int_0^1 (\log \tfrac{1}{t})^{k-1} t^n dt \qquad (k > 0) \quad \text{that} \quad H_{n\nu}^k = \frac{\binom{n}{\nu}}{\Gamma(k)} \int_0^1 (\log \tfrac{1}{t})^{k-1} t^\nu (1-t)^{n-\nu} dt.$$

c) If $d_n = \frac{1}{(p+1)^n}$ then $\Delta d_n = \frac{p}{p+1} d_n$ and

$$\binom{n}{\nu} \Delta^{n-\nu} d_\nu = \binom{n}{\nu} \left(\frac{p}{p+1}\right)^{n-\nu} \frac{1}{(p+1)^\nu} = \frac{1}{(p+1)^n} \binom{n}{\nu} p^{n-\nu} ,$$

i.e., this sequence generates the Euler means.

We now turn to the problem of regularity for Hausdorff means, that is we ask for conditions under which the matrix $a_{n\nu} = \binom{n}{\nu} \Delta^{n-\nu} d_\nu$ satisfies the conditions of the Toeplitz theorem.

It follows from $\sum_{\nu=0}^{n} (AC_1)_{n\nu} = \sum_{\nu=0}^{n} (C_1A)_{n\nu}$, i.e., from

$$\sum_{\nu=0}^{n} a_{n\nu} = \frac{1}{n+1} \sum_{\nu=0}^{n} \sum_{\mu=0}^{\nu} a_{\nu\mu} \,, \quad \text{that} \quad n \sum_{\nu=0}^{n} a_{n\nu} = \sum_{\nu=0}^{n-1} \sum_{\mu=0}^{\nu} a_{\nu\mu} \,.$$

This last formula enables us to determine (by induction) the values of the row sums $\sum_{\nu=0}^{n} a_{n\nu}$. First $a_{00} = d_0$ and we see by induction that $\sum_{\nu=0}^{n} a_{n\nu} = d_0$. Thus, (RS_1) is satisfied iff $d_0 = 1$.

As to (RN) we need the following result (solution of the Hausdorff moment problem, see for instance [32], [25], [68], [106]) : $\sum_{\nu=0}^{n} \binom{n}{\nu} |\Delta^{n-\nu} d_\nu| = O(1)$ holds iff $g(t) \in V(0,1)$ exists such that $d_n = \int_0^1 t^n dg(t)$ (a sequence $\{d_n\}$ of this type is called a moment sequence). Hence, (RN) is satisfied iff $\{d_n\}$ is a moment sequence.

The condition (RN) implies that $\binom{n}{\nu} \Delta^{n-\nu} d_\nu \to 0$ for $n \to \infty$, $\nu \geq 1$, fixed. This can be shown in the following way. Let

$$g^*(t) = \begin{cases} g(t) & t > 0 \\ g(+0) & t = 0 \end{cases} \,, \qquad \text{then for} \qquad \nu \geq 1$$

$$\binom{n}{\nu} \Delta^{n-\nu} d_\nu = \binom{n}{\nu} \int_0^1 t^\nu (1-t)^{n-\nu} dg(t) = \binom{n}{\nu} \int_0^1 t^\nu (1-t)^{n-\nu} dg^*(t) = \binom{n}{\nu} \left(\int_0^\varepsilon + \int_\varepsilon^1 \right) = I + II \,.$$

For $0 \leq t \leq 1$ the function $t^\nu (1-t)^{n-\nu}$ reaches its maximum at $t = \frac{\nu}{n}$, and therefore

$$|I| \leq \binom{n}{\nu} \left(\frac{\nu}{n}\right)^\nu \left(1-\frac{\nu}{n}\right)^{n-\nu} \int_0^\varepsilon |dg^*(t)| = O(1) \int_0^\varepsilon |dg^*(t)| \quad ;$$

here $g^*(t)$ is continuous at the origin and the integral is small for ε small, while $O(1)$ does not depend on ε .

Furthermore, for large n

$$|II| \leq \binom{n}{\nu} \varepsilon^\nu (1-\varepsilon)^{n-\nu} \int_0^1 |dg(t)| \to 0 \qquad \text{as} \qquad n \to \infty \quad .$$

If $\nu = 0$ then

$$\binom{n}{0}\Delta^n d_0 = \int_0^1 (1-t)^n dg(t) = \int_0^1 (1-t)^n dg^*(t) + (g(0)-g^*(0)) .$$

As in the foregoing proof, the integral tends to zero, and (C_0) is satisfied iff $g(0) = g^*(0)$, i.e., iff $g(t)$ is continuous at $t = 0$.

Thus we have the following result: A Hausdorff matrix $\binom{n}{\nu}\Delta^{n-\nu} d_\nu$ is regular iff $d_n = \int_0^1 t^n dg(t)$, where $g(t) \in V(0,1)$, $g(1) - g(0) = 1$, and $g(t)$ is continuous at $t = 0$.

5. Finally we mention a few more summability methods which are not matrix methods.

A sequence $\{s_n\}$ is called limitable by the Abel method A to s if $\Sigma\, x^\nu s_\nu$ exists for $|x| < 1$, and if $\sigma(x) = \sum_{\nu=0}^{\infty} x^\nu s_\nu \,/\, \sum_{\nu=0}^{\infty} x^\nu = (1-x) \sum_{\nu=0}^{\infty} x^\nu s_\nu \to s$ as $x \to 1-0$. If we restrict x to run through values $x_n \to 1$ only, then a matrix method results which is regular by the Toeplitz Theorem. This is true for every such sequence and hence Abel's method itself is regular. (This statement is, of course, the Abel limit theorem.) In the following examples the same reasoning leads to regularity.

Given a sequence $0 \leq \lambda_0 < \lambda_1 < \lambda_2 \ldots \to \infty$ and an order $\kappa \geq 0$, then $s_n = a_0 + \ldots + a_n$ is called limitable to s by the Riesz method (R, λ, κ) (introduced by M. Riesz in [88]) if $\sigma(\omega) = \omega^{-k} \sum_{\lambda_\nu < \omega} (\omega - \lambda_\nu)^\kappa a_\nu \to s$ as $\omega \to \infty$. This method is regular. If $\sigma(\lambda_n) \to s$, then $\{s_n\}$ is called limitable to s by the discontinuous Riesz method (R^*, λ, κ). Obviously, $(R, \lambda, \kappa) \subsetneq (R^*, \lambda, \kappa)$. These methods are equivalent when $0 \leq \kappa \leq 1$ (W. Jurkat [39]); for $\kappa > 1$ there are sequences $\{\lambda_n\}$ such that equivalence does not hold (B. Kuttner [60]), but for "most" sequences $\{\lambda_n\}$ these methods are equivalent also when $1 < \kappa < 2$ ([86]). Furthermore, $(R, \lambda_n = n, \kappa) \approx C_\kappa$ (this was stated in [89] ; proofs may be found in [35], [40], [75]). The Riesz method (R, λ, κ) was introduced in order to study the summability of Dirichlet series $\Sigma\, a_n e^{-\lambda_n s}$.

A sequence $\{s_n\}$ is called limitable to s by Borel's method B if $\Sigma \frac{x^\nu}{\nu!} s_\nu$ exists for all x and if $\sigma(x) = e^{-x} \sum_{\nu=0}^{\infty} \frac{x^\nu}{\nu!} s_\nu \to s$ $(x \to \infty)$. This method is regular.

The following relations hold.

$C_\alpha \subseteqq A$ for every $\alpha \geq 0$ (see G. Frobenius [18]). This can be shown in the following way. If $s_n \to 0$ (C_α) then $s_n = o(n^\alpha)$ and $\sum_{\nu=0}^{\infty} x^\nu s_\nu$ exists for $|x|<1$. Furthermore,

$$\sigma(x) = (1-x)\sum_{\nu=0}^{\infty} x^\nu s_\nu = (1-x)^{\alpha+1}\sum_{\nu=0}^{\infty} x^\nu S_\nu^\alpha = (1-x)^{\alpha+1}\sum_{\nu=0}^{\infty} x^\nu A_\nu^\alpha \sigma_\nu^\alpha .$$

It follows from the Toeplitz Theorem that $(1-x)^{\alpha+1}\sum_{\nu=0}^{\infty} x^\nu A_\nu^\alpha t_\nu$ converges for $x \to 1-0$ when $\{t_\nu\}$ converges, and this shows that $C_\alpha \subseteqq A$.

If $\kappa < \kappa'$, then $(R, \lambda, \kappa) \subseteqq (R, \lambda, \kappa')$. (This is the so called first theorem of consistency for Riesz means; for a proof see [15] , [31].)

$E_p \subseteqq B$ for every $p > 0$ (K. Knopp, [53]). This is a consequence of the identity

$$e^{-x}\sum_{\nu=0}^{\infty} \frac{x^\nu}{\nu!} s_\nu = e^{-(p+1)x}\sum_{\nu=0}^{\infty} \frac{(x(p+1))^\nu}{\nu!}\sigma_\nu ,$$

where $\{\sigma_n\}$ is the E_p transform of $\{s_n\}$.

Problems.

1. Find all (complex) z such that $\{z^n\}$ is limitable by C_α, A, E_p, B, respectively.
2. Show that the product $M_p M_q$ of two arithmetical means is never an arithmetical mean.
3. Show that normal and regular methods M_p, M_q exist such that $M_p \subseteqq M_p M_q$ is not true.
4. Find all sequences that are limitable by N_p, $p_0 = p_1 = 1$, $p_2 = -6$, $p_n = 0$ for $n \geq 3$.
5. Show that $E_p E_q = E_{(p+1)(q+1)-1}$.
6. Show that neither $C_1 \subseteqq E_1$ nor $E_1 \subseteqq C_1$ is true.
7. Show that $\{(n+1)/2^n\}$ is not a moment sequence.
8. Let H be a regular Hausdorff method, $a_{nn} = \int_0^1 t^n dg(t)$. The number $\rho = \inf t$ for all t with $g(x) = g(1)$ for $t \leq x \leq 1$ is called the order of H. Prove the following statements:

   (i) If ρ is the order of H, then $H \supseteqq E_{\frac{1}{\rho}-1}$.

(ii) If ρ is the order of H, then Σz^n is summable to $\frac{1}{1-z}$ for all (complex) z with $\left|z-\frac{\rho-1}{\rho}\right| < \frac{1}{\rho}$.

9. Show that $\Sigma (n+1)^k z^n$, $k = 0, 1, \ldots$, is Abel summable for $|z| \le 1$, $z \ne 1$.

10. Show that $(R, \lambda_n = n, 1) \approx C_1$.

11. Show that $(R, \lambda, 1) \approx (R^*, \lambda, 1) \approx M_p$, $p_n = \lambda_{n+1} - \lambda_n$.

12. Find all (complex) z such that $\Sigma (-1)^n z^{2n}$ is B-summable.

4. Perfect methods.

Definition. A regular, normal matrix $A = (a_{n\nu})$ is called perfect if $\sum_{n=\nu}^{\infty} \alpha_n a_{n\nu} = 0$, $\Sigma|\alpha_n| < \infty$ implies $\alpha_n = 0$, $n = 0, 1, 2, \ldots$ (S. Mazur [70]). In other words, A is perfect if the only vector $\{\alpha_n\} \in \ell_1$ which is orthogonal to all columns of A is the null-vector.
(Note: Our notion of perfecticity of A implies that A is regular and normal.)

Theorem II. 8. Let A be perfect, and $B = (b_{n\nu})$ be triangular. Suppose that (C_0) holds for B, and that $\sigma_n = \sum_{\nu=0}^{n} a_{n\nu} s_\nu = o(1)$ implies $\tau_n = \sum_{\nu=0}^{n} b_{n\nu} s_\nu = O(1)$. Then $\sigma_n = o(1)$ implies $\tau_n = o(1)$.

Proof. By the usual rules for matrix multiplication the transformation $\sigma_n = \sum_{\nu=0}^{n} a_{n\nu} s_\nu$ can be written in the form $\sigma = As$. If $C = BA^{-1}$, then $\tau = Bs = (BA^{-1})\sigma = C\sigma$; hence, there is K with $|c_{n\nu}| \le K$ (by Toeplitz's theorem, C satisfies (RN) since $\sigma_n = o(1)$ is transformed into $\tau_n = O(1)$).

We assume that $\sigma_n \to 0$ exists such that $\tau_n \not\to 0$, i.e., there is some $\delta > 0$ and a sequence $\{n_i\}$ such that $|\tau_{n_i}| \geq \delta$. It follows from $|c_{n\nu}| \leq K$ by the selection principle (see e.g. [106]) that a subsequence $\{m_i\}$ of $\{n_i\}$ exists such that $c_{m_i\nu} \to \alpha_\nu$ $(i \to \infty)$, say, for every fixed ν. Obviously $\Sigma |\alpha_\nu| < K$ because of $\sum_{\nu=0}^{m_i} |c_{m_i\nu}| \leq K$.

But, for $\nu < N < m_i$,

$$\sum_{n=\nu}^{\infty} \alpha_n a_{n\nu} = \sum_{n=\nu}^{m_i} c_{m_i n} a_{n\nu} + \sum_{n=\nu}^{N} (\alpha_n - c_{m_i n}) a_{n\nu} + \sum_{n=N+1}^{m_i} (\alpha_n - c_{m_i n}) a_{n\nu} + \sum_{n=m_i+1}^{\infty} \alpha_n a_{n\nu}$$

$$= I + II + III + IV .$$

First we have $I = b_{m_i\nu} \to 0$ as $i \to \infty$. Given $\varepsilon > 0$, we determine J such that $|IV| \leq \frac{\varepsilon}{4}$ for $i \geq J$. Next, $|III| \leq \sup_{n \geq N+1} |a_{n\nu}| \sum_{n=N+1}^{\infty} (|\alpha_n| + |c_{m_i n}|) \leq \sup_{n \geq N+1} |a_{n\nu}| \cdot 2K \leq \frac{\varepsilon}{4}$ for N sufficiently large and for all i. For a fixed N and all large i (and $i \geq J$) we have $|II| \leq \frac{\varepsilon}{4}$ (because of $c_{m_i n} \to \alpha_n$ as $i \to \infty$). It follows that $\sum_{n=\nu}^{\infty} \alpha_n a_{n\nu} = 0$, this implies $\alpha_n = 0$ and hence $(c_{m_i n})$ satisfies (C_0). (Note that $\alpha_n = \lim_i c_{m_i n}$.) But this yields $\tau_{m_i} \to 0$ in contradiction to our assumption that $|\tau_{n_i}| \geq \delta > 0$.

<u>Theorem II. 9.</u> Let A be perfect. If $G_n = \sum_{\nu=0}^{n} |a'_{n\nu}|$, and if $s_n \to 0$ (A), then $s_n = o(G_n)$ (and this estimate is best possible by Theorem II. 3.).

<u>Proof.</u> Apply Theorems II. 3. and II. 8. to A and $B = (\delta_{n\nu} G_n^{-1})$.

Next we will discuss two sufficient criteria for the perfecticity of a matrix method.

<u>Lemma II. 2.</u> A normal and regular method $A = (a_{n\nu})$ is perfect, if $\sup_n |a'_{n\nu}| < \infty$ for every $\nu = 0, 1, 2 \ldots$.

<u>Proof.</u> $\Sigma |\alpha_n| < \infty$, $\sum_{n=\nu}^{\infty} \alpha_n a_{n\nu} = 0$ imply

$$0 = \sum_{\nu=k}^{\infty} a'_{\nu k} \sum_{n=\nu}^{\infty} \alpha_n a_{n\nu} = \sum_{n=k}^{\infty} \alpha_n \sum_{\nu=k}^{n} a_{n\nu} a'_{\nu k} = \alpha_k .$$

It follows from Lemma II. 2, that all Cesàro methods C_α are perfect; in this case $a'_{n\nu} = A_{n-\nu}^{-\alpha-1} A_\nu^\alpha \to 0$ for $n \to \infty$. Also all weighted arithmetical means are perfect; in this case $a'_{n\nu} = 0$ for $\nu \leq n-2$.

On the other hand, not all Nörlund means are perfect. In the case of $\sigma_n = 2s_{n-1} - s_n$, this follows from Theorem II. 9. More generally, a Nörlund method N_p is not perfect if some ξ, $|\xi| < 1$, is a zero of $p(z) = \sum_0^\infty p_n z^n$. If we write $\alpha_n = P_n \xi^n$ in this case, then $\Sigma |\alpha_n| < \infty$ (since N_p is regular) and $\sum_{n=\nu}^\infty \alpha_n \frac{p_{n-\nu}}{P_n} = \xi^\nu p(\xi) = 0$. The question arises whether a nonperfect method N_p exists such that $p(z) \neq 0$ for $|z| < 1$.

Not every Hausdorff method is perfect. We mention the Haudorff method A with $d_n = \frac{3}{n+1} - 2$, i.e., $A = 3C_1 - 2C_0$. It is not difficult to see that $s_n = \sqrt{n+1}$ is limitable by this method. On the other hand,

$$a'_{n\nu} = -\frac{1}{2}\,\delta_{n\nu} - \frac{3}{4}\binom{n}{\nu}\frac{\Gamma(\nu-\frac{1}{2})\,\Gamma(n-\nu+1)}{\Gamma(n+\frac{1}{2})}$$

and

$$G_n = \sum_{\nu=0}^{n} |a'_{n\nu}| = a'_{no} - \sum_{\nu=1}^{n} a'_{n\nu} = 2a'_{no} - 1 = 3\,\frac{\Gamma(\frac{1}{2})\Gamma(n+1)}{\Gamma(n+\frac{1}{2})} - 1 \sim \sqrt{n}\,.$$

It follows from Theorem II. 9. that this method is not perfect. In this case we also can give explicitly a sequence $\{\alpha_n\}$, $\alpha_n \not\equiv 0$, $\Sigma |\alpha_n| < \infty$ such that $\sum_{n=\nu}^\infty \alpha_n a_{n\nu} = 0$, for instance

$$\alpha_n = \prod_{\nu=0}^{n-1} \frac{\nu-\frac{1}{2}}{\nu+1} \sim n^{-\frac{3}{2}}\,, \qquad \alpha_0 = 1\,.$$

One can show that certain properties of the so called <u>moment function</u> $\mu(z) = \int_0^1 t^z dg(t)$ (in particular the location of its zeros) are important to determine what sequences are limitable by the corresponding Hausdorff method. For Nörlund methods, $p(z)$ plays a similar role (see chapter IV, 3; for further information see e.g. [75], [2], [36]).

<u>Lemma II. 3.</u> A normal and regular method $A = (a_{n\nu})$ is perfect if a sequence $\{z_i\}$, $\sup |z_i| < 1$, $z_i \neq z_j$ for $i \neq j$, exist such that

$$z_i^n = \sum_{\nu=0}^{n} a_{n\nu} s_\nu(z_i)\,, \qquad s_\nu(z_i) = O(1) \qquad (\nu \to \infty\,,\ i \text{ fixed})\,.$$

Proof. $\Sigma |\alpha_n| < \infty$, $\sum_{n=\nu}^{\infty} \alpha_n a_{n\nu} = 0$ imply

$$\sum_{n=0}^{\infty} \alpha_n z_1^n = \sum_{n=0}^{\infty} \alpha_n \sum_{\nu=0}^{n} a_{n\nu} s_\nu(z_1) = \sum_{\nu=0}^{\infty} s_\nu(z_1) \sum_{n=\nu}^{\infty} \alpha_n a_{n\nu} = 0 ,$$

and it follows (uniqueness of the coefficients of a power series) that $\alpha_n = 0$, $n = 0, 1, 2 \dots$.

By Lemma II. 3. we show that all Euler methods are perfect. In this case

$$\sum_{\nu=0}^{n} a'_{n\nu} z^\nu = \sum_{\nu=0}^{n} \binom{n}{\nu}(-1)^{n-\nu} p^{n-\nu} (z(p+1))^\nu = (z(p+1)-p)^n ,$$

are bounded functions $s_n(z)$ for z with $|z - \frac{p}{p+1}| < \frac{1}{p+1}$.

Theorem II. 10. (J. D. Hill [33]) Let A and B be perfect. Then AB is perfect.

Proof. Let α be a vector $(\alpha_0, \alpha_1, \alpha_2, \dots)$, $\Sigma |\alpha_n| < \infty$, and assume that $0 = \alpha(AB) = (\alpha A)B = \beta B$. We have $\sum_\nu |\beta_\nu| \leq \sum_\nu \sum_n |\alpha_n a_{n\nu}| = \sum_n \alpha_n \sum_\nu |a_{n\nu}| < \infty$.
B is perfect which implies that β is the null vector, and, because A is perfect, α is the null vector.

If $A \approx B$ and if B is perfect, then A is perfect. In this case, $A = CB$, where C and C^{-1} are regular. Thus, C is perfect by Lemma II. 2. and A is perfect by Theorem II. 10.

By the theorem of Knopp and Schnee ($C_\alpha \approx H^\alpha$; it will be proven in section 7 of this chapter) and by this remark we obtain that the methods H^α are perfect. (For further results about perfect methods see [33], [87] .)

Theorem II. 11. (S. Mazur and W. Orlicz [71]) Let A be normal and regular, and suppose that only bounded sequences are limitable by A . Then, A is equivalent to convergence.

Proof. If $\sigma_n \to 0$, then $s_n = \sum_{\nu=0}^{n} a'_{n\nu} \sigma_\nu = O(1)$, and this implies $G_n = O(1)$ by Theorem II. 1. . Therefore, A is perfect by Lemma II. 2. and $s_n = o(1)$ by Theorem II. 9. .
(For generalizations of Theorem II. 11. see [72] , [34], [109], [110] .)

Theorem II. 12. (S. Mazur [70] see also [72]) Let A be perfect. Suppose that B is triangular and regular, and that every sequence $s_n \to 0$ (A) is limitable by the method B . Then A and B are consistent.

Proof. This is an immediate consequence of Theorem II. 8.. (S. Mazur introduced the notion of perfecticity in connection with this theorem.)

Theorem II. 12. shows that perfect methods are consistent with stronger methods. In this theorem perfecticity cannot be omitted. This is shown by the example A: $\sigma_n = 2s_{n-1} - s_n$, B: $\tau_n = 2s_{n-1} - s_n + \frac{s_n}{2^n}$; it follows from earlier results on Nörlund methods (see p. 18) that $A \subsetneqq B$. But $s_n = 2^n \to 0$ (A) , $s_n \to 1$ (B) .

We finally show that every method which is consistent with every stronger method is perfect.

Theorem II. 13. (S. Banach [5]) A normal and regular matrix A is perfect if it is consistent with every triangular and regular matrix $B \supseteqq A$.

Proof. Let A (normal and regular) be not perfect. We show that $B \supseteqq A$ exists that it is not consistent with A .

There is $\alpha_n \not\equiv 0$, $\Sigma |\alpha_n| < \infty$, $\sum_{n=\nu}^{\infty} \alpha_n a_{n\nu} = 0$ ($\nu = 0, 1, 2, \ldots$) . We write $A_n = \sum_{\nu=0}^{n} a_{n\nu}$, $\sum_{0}^{\infty} \alpha_n A_n = \alpha$ and define a matrix B by $b_{n\nu} = (1-\alpha)a_{n\nu} + \sum_{\mu=\nu}^{n} \alpha_\mu a_{\mu\nu}$ $(\nu \leq n)$, $b_{n\nu} = 0$ $(\nu > n)$.

The matrix $B = (b_{n\nu})$ is regular, since

$$\sum_{\nu=0}^{n} |b_{n\nu}| \leq |1-\alpha| \sum_{\nu=0}^{n} |a_{n\nu}| + \sum_{\mu=0}^{n} |\alpha_\mu| \sum_{\nu=0}^{\mu} |a_{\mu\nu}| \leq K ,$$

$$\sum_{\nu=0}^{n} b_{n\nu} = (1-\alpha) A_n + \sum_{\mu=0}^{n} \alpha_\mu A_\mu \to 1 ,$$

$$b_{n\nu} = o(1) + \sum_{\mu=\nu}^{n} \alpha_\mu a_{\mu\nu} \to \sum_{\mu=\nu}^{\infty} \alpha_\mu a_{\mu\nu} = 0 \qquad (n \to \infty) .$$

Furthermore, $A \subseteqq B$ since $s_n \to s$ (A) implies

$$\sum_{\nu=0}^{n} b_{n\nu} s_\nu = (1-\alpha)\sigma_n + \sum_{\mu=0}^{n} \alpha_\mu \sigma_\mu \to (1-\alpha)s + \sum_{\mu=0}^{\infty} \alpha_\mu \sigma_\mu .$$

We may choose $\{\sigma_n\}$ such that $\sigma_n \to 0$, $\sum_{\mu=0}^{\infty} \alpha_\mu \sigma_\mu \nrightarrow 0$, which shows that B is not consistent with A.

Problems.

1. Show that $\alpha C_0 + (1-\alpha) C_1$ is perfect iff $\alpha \geq 0$.
2. Show that $A = (a_{n\nu})$ with $a_{n\nu} = \frac{1}{n+\alpha}$ $(\nu < n)$, $a_{nn} = \frac{\alpha}{n+\alpha}$, $a_{n\nu} = 0$ $(\nu > n)$ is perfect for $\alpha \geq \frac{1}{2}$ and is not perfect for $0 < \alpha < \frac{1}{2}$. (In this connection see [47], [86], [73].)
3. Find all values of α such that $\alpha E_0 + (1-\alpha) E_1$ is perfect.
4. Let A and B be normal and regular. Show that A is perfect if AB is perfect.

5. Mean value conditions

A triangular matrix $A = (a_{n\nu})$ satisfies the mean value condition $M_K(A)$ if

$$\left|\sum_{\nu=0}^{m} a_{n\nu} s_\nu\right| \leq K \sup_{\mu \leq m} |\sigma_\mu| \qquad (m \leq n,\ K \text{ independent of } m, n \text{ and } \{s_\nu\})$$

(see [31], Lemma 8, [9], [81], [111]). For instance, $M_1(M_p)$ is true when $p_n \geq 0$, since in this case

$$\left|\frac{1}{P_n} \sum_{\nu=0}^{m} p_\nu s_\nu\right| = \frac{P_m}{P_n} |\sigma_m| \leq |\sigma_m| .$$

In particular, $M_1(C_1)$ is true.

We give necessary and sufficient conditions for $M_K(A)$. In case $a_{nn} \neq 0$ we have

$$\sum_{\nu=0}^{m} a_{n\nu} s_\nu = \sum_{\nu=0}^{m} a_{n\nu} \sum_{\mu=0}^{\nu} a'_{\nu\mu} \sigma_\mu = \sum_{\mu=0}^{m} \sigma_\mu \sum_{\nu=\mu}^{m} a_{n\nu} a'_{\nu\mu} .$$

Then

$$\text{(8)} \qquad \sum_{\mu=0}^{m} \left| \sum_{\nu=\mu}^{m} a_{n\nu} a'_{\nu\mu} \right| \leq K \qquad (m \leq n)$$

is sufficient for $M_K(A)$. It is readily seen that (8) is also necessary (take $\sigma_\mu = \pm 1$, $\mu = 0, \ldots, m$, appropriately; m, n fixed).

Lemma II. 4. A normal and regular matrix A that satisfies $M_K(A)$ is perfect.

Proof. Let $\Sigma |\alpha_n| < \infty$, $\sum_{n=\nu}^{\infty} \alpha_n a_{n\nu} = 0$. Then for every $k \geq \mu$ we have

$$0 = \sum_{\nu=\mu}^{k} a'_{\nu\mu} \sum_{n=\nu}^{\infty} \alpha_n a_{n\nu} = \sum_{n=\mu}^{k} \alpha_n \delta_{n\mu} + \sum_{n=k+1}^{\infty} \alpha_n \sum_{\nu=\mu}^{k} a_{n\nu} a'_{\nu\mu} = \alpha_\mu + \sum_{n=k+1}^{\infty} \alpha_n O(1)$$

because of (8), and hence $\alpha_\mu = 0$.

Theorem II. 14. (cf. [43]) Let A be normal and regular; suppose that $M_K(A)$ holds. Then $s_n \to s$ (A) implies $s_n = s + o(\frac{1}{a_{nn}})$.

Proof. We have $a_{nn} s_n = \sigma_n - \sum_{\nu=0}^{n-1} a_{n\nu} s_\nu = O(1)$, and it follows from Lemma II. 4. and Theorem II. 8. that $a_{nn}(s_n - s) = o(1)$.

As a consequence $M_K(E_1)$ does not hold. Here, $s_n = o(3^n)$ is the best estimate, whereas $s_n = o(2^n)$ would follow from Theorem II. 14..

Theorem II. 15. ([47]) Let A be triangular and regular; suppose that $M_K(A)$ holds. Then $\Sigma |a_{nn}| = \infty$.

Proof. The regularity of A implies $\sum_\nu \sup_n |a_{n\nu}| = \infty$. Otherwise

$$\left| \sum_{\nu=0}^{n} a_{n\nu} \right| \leq \sum_{\nu=0}^{N} |a_{n\nu}| + \sum_{\nu=N+1}^{\infty} \sup_n |a_{n\nu}| \leq \varepsilon$$

Furthermore, it follows from $M_K(A)$ applied to $s_n = \delta_{mn}$ that

$$\text{(9)} \qquad |a_{nm}| \leq K |a_{mm}| .$$

This is $\sup_n |a_{n\nu}| \leq K |a_{\nu\nu}|$, so that $\sum_\nu |a_{\nu\nu}| = \infty$.

It follows from Theorem II. 15. that $M_K(C_\alpha)$ is not true for $\alpha > 1$ (in this case $a_{nn} \sim \frac{1}{n^\alpha}$).

The series Σz^n cannot be summable for any $|z| > 1$ by a matrix A which is normal, regular and satisfies $M_K(A)$. If it were, then it follows $\left|\frac{1-z^{n+1}}{1-z}\right| = O(\frac{1}{a_{nn}})$, that is $a_{nn} = O(\frac{1}{1-z^{n+1}})$ by Theorem II. 14., so that we have $\Sigma |a_{nn}| < \infty$ in violation of Theorem II. 15. .

Next we discuss some sufficient conditions for $M_K(A)$.

Lemma II. 5. Let $A = (a_{n\nu})$ be normal. Assume that $a_{n\nu} > 0$ $(\nu \leq n)$ and that

(10) $$\frac{a_{n+1,\nu}}{a_{n\nu}} \downarrow \quad \text{for} \quad \nu \uparrow \quad (\nu \leq n) .$$

Then $a'_{n\nu} \leq 0$ for $\nu < n$.
(For a sharper result see [84].)

Proof. First, $a'_{10} = - \frac{a_{10}}{a_{11}a_{00}} < 0$.

We proceed by induction and assume that $a'_{\mu\nu} \leq 0$ for $\mu \leq n$, $\nu < \mu$. We have for every $\nu \leq n$

$$0 = \sum_{\mu=\nu}^{n+1} a_{n+1,\mu}a'_{\mu\nu} = a_{n+1,n+1}a'_{n+1,\nu} + \sum_{\mu=\nu}^{n} \left(\frac{a_{n+1,\mu}}{a_{n\mu}} - \frac{a_{n+1,\nu}}{a_{n\nu}}\right)a_{n\mu}a'_{\mu\nu} + \frac{a_{n+1,\nu}}{a_{n\nu}}\delta_{n\nu}$$
$$\geq a_{n+1,n+1}a'_{n+1,\nu} .$$

Theorem II. 16. (see [44]) Let A be normal, and suppose that $a'_{n\nu} \leq 0$ $(\nu < n)$, $a_{nn} > 0$.
Then,

(i) $$a_{n\nu} \geq 0 \quad \text{for} \quad \nu \leq n ,$$

(ii) $$M_1 \quad \text{holds if} \quad \sum_{\nu=0}^{n} a'_{n\nu} \geq 0 \qquad (n \geq 0) ,$$

(iii) $$\left|\sum_{\nu=0}^{m} a_{n\nu}s_\nu\right| \leq \left(\sum_{\nu=0}^{m} a_{n\nu}\right) \sup_{\mu \leq m} \frac{|\sigma_\mu|}{A_\mu} , \qquad A_n = \sum_{\nu=0}^{n} a_{n\nu} .$$

Proof. Induction yields (i) : Let $a_{n\mu} \geq 0$ for $\nu + 1 \leq \mu \leq n$ (n fixed), then

$$a_{n\nu}a'_{\nu\nu} \geq a_{n\nu}a'_{\nu\nu} + \sum_{\mu=\nu+1}^{n} a_{n\mu}a'_{\mu\nu} = 0 .$$

Next, we turn to the left-hand side of (8) . We have for $\mu \leq m < n$

$$\sum_{\nu=\mu}^{m} a_{n\nu}a'_{\nu\mu} = -\sum_{\nu=m+1}^{n} a_{n\nu}a'_{\nu\mu} \geq 0$$

(because of $a_{n\nu}a'_{\nu\mu} \leq 0$ for $\mu < \nu$) and therefore

$$(11) \qquad \sum_{\mu=0}^{m} \left| \sum_{\nu=\mu}^{m} a_{n\nu}a'_{\nu\mu} \right| = \sum_{\nu=0}^{m} a_{n\nu} \sum_{\mu=0}^{\nu} a'_{\nu\mu} \qquad (m \leq n) .$$

If $\sum_{\mu=0}^{\nu} a'_{\nu\mu} \geq 0$, then the right-hand side of (11) is at most $\sum_{\nu=0}^{n} a_{n\nu} \sum_{\mu=0}^{\nu} a'_{\nu\mu} = \sum_{\mu=0}^{n} \delta_{n\mu} = 1$, which proves (ii).

In order to prove (iii) we proceed as follows. Corresponding to (11) the equality

$$\sum_{\mu=0}^{m} \left| \sum_{\nu=\mu}^{m} a^*_{n\nu}a^{*\prime}_{\nu\mu} \right| = \sum_{\nu=0}^{m} a^*_{n\nu} \quad , \quad A^* = (a^*_{n\nu}) = \left(\frac{a_{n\nu}}{A_n}\right) ,$$

(note that $a^{*\prime}_{n\nu} = a'_{n\nu} A_\nu$ and $\sum_{\nu=0}^{\nu} a^{*\prime}_{\nu\mu} = 1$) holds, and the proof of (8) - applied to A^* - shows that (iii) is true.

Theorem II. 16. - possibly in combination with Lemma II. 5. - gives sufficient conditions for $M_K(A)$. We will postpone the discussion of examples a little and first examine the assumptions of Theorem II. 16.. Let $A = (a_{n\nu})$ be normal and assume that (i) of Theorem II. 16. holds. We will say that A satisfies the <u>mean value condition $M^*_K(A)$</u> if

$$(12) \qquad \left| \sum_{\nu=0}^{m} a_{n\nu}a_\nu \right| \leq K \left(\sum_{\nu=0}^{m} a_{n\nu} \right) \sup_{\mu \leq m} \frac{|\sigma_\mu|}{A_\mu} \quad , \quad A_n = \sum_{\nu=0}^{n} a_{n\nu} ,$$

($m \leq n$, K independent of m, n and $\{s_\nu\}$).

<u>Theorem II. 17.</u> ([47], [113]) If $M^*_1(A)$ holds, then $a'_{n\nu} \leq 0$ for $\nu < n$.

<u>Proof.</u> Clearly, we may assume that $A_n = 1$. It follows from the identity

$$\sum_{\nu=0}^{n-1} a'_{n\nu}\sigma_\nu = s_n - \frac{\sigma_n}{a_{nn}} = -\frac{1}{a_{nn}} \sum_{\nu=0}^{n-1} a_{n\nu}s_\nu$$

and $M^*_1(A)$ that

$$\sum_{\nu=0}^{n-1} |a'_{n\nu}| \leq \frac{1}{a_{nn}} \sum_{\nu=0}^{n-1} a_{n\nu} = \frac{1}{a_{nn}} - 1 \qquad (\text{take } \sigma_\nu = \pm 1) ,$$

hence

$$\sum_{\nu=0}^{n-1} (|a'_{n\nu}| + a'_{n\nu}) \leq \frac{1}{a_{nn}} - 1 + (1 - a'_{nn}) = 0 .$$

But $|a'_{n\nu}| + a'_{n\nu} \geq 0$, and this shows that $a'_{n\nu} \leq 0$ for $\nu < n$.

We discuss some examples. Let $A = C_\alpha$, $0 \le \alpha \le 1$, i.e. $a_{n\nu} = A_{n-\nu}^{\alpha-1} / A_n^\alpha$. It follows from $a'_{n\nu} = A_{n-\nu}^{-\alpha-1} A_\nu^\alpha \le 0$ $(\nu < n)$ and Theorem II. 16. that $M_1^*(C_\alpha)$ and $M_1(C_\alpha)$ hold when $0 \le \alpha \le 1$ ([9], [13]). If $a_{n\nu} = A_{n-\nu}^{\alpha-1}$, $0 \le \alpha \le 1$, then $a'_{n\nu} = A_{n-\nu}^{-\alpha-1}$, $\sum_{\nu=0}^{n} a'_{n\nu} = A_n^{-\alpha} \ge 0$, and we obtain, from Theorem II. 16. (ii), the inequality

$$(13) \qquad \left| \sum_{\nu=0}^{m} A_{n-\nu}^{\alpha-1} s_\nu \right| \le \sup_{\mu \le m} \left| \sum_{\nu=0}^{\mu} A_{\mu-\nu}^{\alpha-1} s_\nu \right|$$

which will be used in chapter III .

For a regular and positive Nörlund mean N_p it follows from Lemma II. 5. and Theorem II. 16. that $M_1^*(N_p)$ and $M_1(N_p)$ is satisfied if

$$(14) \qquad \frac{p_{\nu+1}}{p_\nu} \uparrow \quad \text{as} \quad \nu \uparrow \quad .$$

We note that, for N_p regular and positive, (14) implies $0 < p_n \downarrow$. This is easily seen from the observation that $p_{n+1} \ge p_n(1+\varepsilon)$ for some $\varepsilon > 0$ and all large n would violate the condition $p_n = o(P_n)$ which is necessary for regularity.

Theorem II. 18. Let A and B be normal. If $A \approx B$ and if $M_K(A)$ holds, then $M_L(B)$ is true for some L .

Proof. $A \approx B$ implies $B = CA$ with C, C^{-1} regular. Using the notation $\sigma_n^A = \sum_{\nu=0}^{n} a_{n\nu} s_\nu$ and similarly σ_n^B, it follows that

$$|\sigma_n^A| = \left| \sum_{\nu=0}^{n} c'_{n\nu} \sigma_\nu^B \right| \le K_1 \sup_{\nu \le n} |\sigma_\nu^B| \quad .$$

But (for $m \le n$)

$$\left| \sum_{\nu=0}^{m} b_{n\nu} s_\nu \right| = \left| \sum_{\nu=0}^{m} s_\nu \sum_{\mu=\nu}^{n} c_{n\mu} a_{\mu\nu} \right| \le \left| \sum_{\mu=0}^{m} c_{n\mu} \sigma_\mu^A \right| + \left| \sum_{\mu=m+1}^{n} c_{n\mu} \sum_{\nu=0}^{m} a_{\mu\nu} s_\nu \right|$$

$$\le \sup_{\nu \le m} |\sigma_\nu^A| \sum_{\mu=0}^{m} |c_{n\mu}| + K \sup_{\nu \le m} |\sigma_\nu^A| \sum_{\mu=m+1}^{n} |c_{n\mu}| \le K_2 \sup_{\nu \le m} |\sigma_\nu^A| \quad .$$

Problems.

1. Show that $0 < \liminf |a_{nn}| G_n \le \limsup |a_{nn}| G_n < \infty$ under the assumptions of Theorem II. 14. .
2. Let A be a normal matrix that satisfies (10), and let $b_n \ge 0$. Let a normal matrix $A_b = (\alpha_{n\nu})$ be defined by $\alpha_{n\nu} = a_{n\nu}(A_n+b_n)^{-1}$ $(\nu < n)$, $\alpha_{nn} = (a_{nn}+b_n)(A_n+b_n)^{-1}$. Show that $M_1^*(A_b)$.

3. Let A be a matrix that satisfies the assumptions of problem 2. , and let A also be regular. Show that $A \subseteqq A_b$ if $b_n = O(a_{nn})$, and that $A_b \subseteqq A$ if $a_{nn} = O(b_n)$, $b_n = O(1)$. As an application show that the matrix A of problem 2., page 31 , is equivalent to C_1 for $\alpha \geq 1$.

4. Show that $\alpha C_0 + (1-\alpha) C_1 \approx C_0$ for $0 < \alpha \leq 1$.

5. Let A, B be normal and regular matrices. Show that $A \subseteqq B$, $M_K(B)$ does not imply $M_L(A)$ for any L.

6. Applications of mean value conditions

We discuss three different applications: Comparison theorems, summability factors, and Tauberian Theorems.

1. Comparison theorems.

Theorem II. 19. ([44]) Assume that A is regular, $a_{n\nu} > 0$ for $\nu \leq n$, and that $M_K(A)$ holds. Let B be triangular and satisfy (C_0) . Then $s_n \to 0$ (A) implies $s_n \to 0$ (B) if

$$\sum_{\nu=0}^{n} \left|\Delta_\nu \frac{b_{n\nu}}{a_{n\nu}}\right| = O(1) \qquad (\Delta_\nu e_{n\nu} = e_{n\nu} - e_{n,\nu+1} \;, \quad \Delta_n \frac{b_{nn}}{a_{nn}} = \frac{b_{nn}}{a_{nn}}) \;.$$

Proof. Let $\sigma_n = \sum_{\nu=0}^{n} a_{n\nu} s_\nu$. By partial summation and $M_K(A)$

$$\tau_n = \sum_{\nu=0}^{n} b_{n\nu} s_\nu = \sum_{\nu=0}^{n} \frac{b_{n\nu}}{a_{n\nu}} a_{n\nu} s_\nu = \sum_{\nu=0}^{n} \Delta \frac{b_{n\nu}}{a_{n\nu}} \sum_{\mu=0}^{\nu} a_{n\mu} s_\mu = \sup_{\nu \leq n} |\sigma_\nu| \, O(1) \quad ,$$

and $\tau_n = o(1)$ because of $\sigma_n = o(1)$ and the perfecticity of A.

Theorem II. 20. ([44]) Assume that A is regular, $a_{n\nu} > 0$ for $\nu \leq n$, and that $M_K^*(A)$ holds. Let B be triangular and regular.

Then $s_n \to 0$ (A) implies $s_n \to 0$ (B) if

$$\sum_{\mu=\nu}^{n} \left|\Delta_\nu \frac{b_{n\mu}}{a_{n\mu}}\right| = O\left(\frac{b_{n\nu}}{a_{n\nu}}\right) + O(1)$$

(uniformly in ν and n).

Proof. As in the proof of Theorem II. 19.,

$$\tau_n = \sup_{\nu \le n} \frac{|\sigma_\nu|}{A_\nu} O(1) \sum_{\nu=0}^{n} \left|\Delta \frac{b_{n\nu}}{a_{n\nu}}\right| \sum_{\mu=0}^{\nu} a_{n\mu} = O(1) \sum_{\mu=0}^{n} a_{n\mu}\left(O\left(\frac{b_{n\mu}}{a_{n\mu}}\right) + O(1)\right) = O(1) .$$

Theorem II. 21. ([47]) Let A and B be normal and regular, and assume that $M_1^*(B)$ holds. Then $A \subseteqq B$ if $0 \le \frac{a_{n\nu}}{b_{n\nu}} \uparrow$ for $\nu \uparrow$ ($\nu \le n$) .

Proof. There is no loss of generality if we assume that $\sum_{\nu=0}^{n} b_{n\nu} = \sum_{\nu=0}^{n} a_{n\nu} = 1$ (n = 0, 1, ...). We show first that $(AB')_{n\nu} \le 0$ for $\nu < n$. In fact,

$$(AB')_{n\nu} = \sum_{\mu=\nu}^{n} \left(\frac{a_{n\mu}}{b_{n\mu}} - \frac{a_{n\nu}}{b_{n\nu}}\right) b_{n\mu} b'_{\mu\nu} + \frac{a_{n\nu}}{b_{n\nu}} \delta_{n\nu} \le 0 \qquad (\nu < n)$$

($b'_{n\nu} \le 0$ for $\nu < n$ follows from Theorem II. 17.). It follows from Theorem II. 16. that $(BA')_{n\nu} \ge 0$ for $\nu \le n$. BA' satisfies conditions (RN), (RS_1) of Theorem II. 1. since

$$\sum_{\nu=0}^{n} (BA')_{n\nu} = \sum_{\mu=0}^{n} b_{n\mu} \sum_{\nu=0}^{\mu} a_{\mu\nu} = \sum_{\mu=0}^{n} b_{n\mu} = 1 .$$

We finally have $(BA')_{n\nu} = \sum_{\mu=\nu}^{n} b_{n\mu} a'_{\mu\nu} \to 0$ for $n \to \infty$, ν fixed, and it follows from Theorem II. 4. that $A \subseteqq B$.

Roughly speaking, Theorems II. 19. - II. 21. require that $\frac{b_{n\nu}}{a_{n\nu}}$ is either monotone decreasing or monotone increasing and bounded, and $A \subseteqq B$ follows if $M_1^*(A)$ or $M_1^*(B)$ is satisfied.

Examples. We have $M_p \subseteqq M_q$ ($p_n > 0$, $q_n > 0$) if $\frac{q_\nu}{p_\nu}$ is decreasing. Roughly speaking: The method M_p becomes weaker the faster $\{p_n\}$ increases. (This result is related to the socalled second theorem of consistency for Riesz means, see [15]. Compare this result also with the results on page 17 .)

It follows from $\frac{(N_q)_{n\nu}}{(N_p)_{n\nu}} = \frac{q_{n-\nu}}{p_{n-\nu}} \frac{P_n}{Q_n}$ that $N_p \subseteqq N_q$ (N_p, N_q regular and positive) if $\frac{q_n}{p_n}\uparrow$ for $n\uparrow$, and if $M_1^*(N_p)$ or $M_1^*(N_q)$ is satisfied. Roughly speaking: The method N_p becomes stronger the faster p_n increases. In particular, it follows that $C_1 \subsetneqq N_p$ if $p_n \uparrow$ and $N_p \subsetneqq C_1$ if $p_n \downarrow$.

Theorem II. 22. $C_\alpha C_1 \approx C_{\alpha+1}$ for $\alpha > -1$.

Proof. A short calculation shows that $C_{\alpha+1} C_\alpha^{-1} = M_p$, $p_n = A_n^\alpha$. Here $\frac{(M_p)_{n\nu}}{(C_1)_{n\nu}} = \frac{(n+1) A_\nu^\alpha}{A_n^{\alpha+1}}$, and it follows from Theorem II. 20. that $M_p \approx C_1$, which yields the statement of the theorem (note that $A \approx B$ implies $AC \approx BC$ if A, B, C are normal and regular, and that all methods involved in the proof are Hausdorff methods).
(For further examples see [44] .)

2. Convergence and summability factors.

The problems discussed here arise from a well-known convergence-criterion (DuBois-Reymond and Dedekind) : $\Sigma\, a_\nu b_\nu$ is convergent if $\Sigma\, a_n$ converges and if $\Sigma |\Delta b_n| < \infty$ (see [56]) . Replacing convergence by summability leads to the following problem:

Given methods A and B , find necessary and sufficient conditions for a sequence $\{\bar{\varepsilon}_n\}$ to have the property

(i) $\Sigma\, a_n \in A$ implies $\Sigma\, a_n \bar{\varepsilon}_n \in B$.

(In what follows a notation $s_n \in A$ resp. $\Sigma\, a_n \in A$ means limitability resp. summability by the method A.)

A similar problem is the following:

Given methods A and B , find necessary and sufficient conditions for a sequence $\{\varepsilon_n\}$, to have the property

(ii) $s_n \in A$ implies $\Sigma\, s_n \varepsilon_n \in B$.

Sequences with the properties (i) resp. (ii) are called summability factors (convergence factors if $B = (\delta_{n\nu})$). It turns out that problem (ii) is easier to solve, and we will discuss (ii) first. Before we go into details, we must introduce a new denotation. Given a triangular matrix $A = (a_{n\nu})$ and a sequence $\{s_n\}$, $s_n = a_0 + \ldots + a_n$, we write

$$\sigma_n = \sum_{\nu=0}^{n} a_{n\nu} s_\nu = \sum_{\nu=0}^{n} a_{n\nu} \sum_{\mu=0}^{\nu} a_\mu = \sum_{\mu=0}^{n} a_\mu \sum_{\nu=\mu}^{n} a_{n\nu} = \sum_{\mu=0}^{n} \bar{a}_{n\mu} a_\mu$$

(note that $a_{nn} = \bar{a}_{nn}$). By the transformation $\sigma_n = \sum_{\mu=0}^{n} \bar{a}_{n\mu} a_\mu$ the terms of a series Σa_n are transformed into a sequence $\{\sigma_n\}$. If A is regular, then $\bar{a}_{n\nu} \to 1$ for $n \to \infty$, ν fixed (by Theorem II. 1.).

Theorem II. 23. (see [81]) Let A and B be normal and regular. Then the factors $\bar{\varepsilon}_n$ in problem (i) necessarily have a representation

(15) $$\bar{\varepsilon}_n = \alpha + \sum_{k=n}^{\infty} \alpha_k \bar{a}_{kn} \qquad \text{for some} \qquad \{\alpha_n\}, \qquad \Sigma|\alpha_n| < \infty$$

and

(16) $$\bar{\varepsilon}_n = O\left(\frac{a_{nn}}{b_{nn}}\right) \quad .$$

Likewise, for (ii) the following conditions are necessary:

(17) $$\varepsilon_n = \sum_{k=n}^{\infty} \alpha_k a_{kn} \qquad \text{for some} \qquad \{\alpha_n\}, \qquad \Sigma|\alpha_n| < \infty$$

and

(18) $$\varepsilon_n = O\left(\frac{a_{nn}}{b_{nn}}\right) \quad .$$

Proof. In order to prove (17) we assume that $\{\tau_n = \sum_{\nu=0}^{n} \bar{b}_{n\nu} \varepsilon_\nu s_\nu\}$ converges whenever $\{\sigma_n = \sum_{\nu=0}^{n} a_{n\nu} s_\nu\}$ converges, i.e.,

$$\tau_n = \sum_{\mu=0}^{n} \sigma_\mu \sum_{\nu=\mu}^{n} \bar{b}_{n\nu} \varepsilon_\nu a'_{\nu\mu} = \sum_{\mu=0}^{n} c_{n\mu} \sigma_\mu$$

is a transformation which is convergence-preserving.

It follows from Theorem II. 1., case 3°, that

$$\lim \tau_n = (\alpha - \Sigma\,\alpha_\nu)\lim_n \sigma_n + \Sigma\,\alpha_\nu\sigma_\nu \qquad (\alpha = \lim \sum_{\nu=0}^{n} c_{n\nu}\,,\ \alpha_\nu = \lim_n c_{n\nu}) \quad .$$

Choosing $s_\nu = \delta_{i\nu}$, then $\sigma_n = a_{ni} \to 0$ for $n \to \infty$ and it follows because of $\bar{b}_{ni} \to 1$ for $n \to \infty$ that

$$\varepsilon_i = \sum_{\nu=i}^{\infty} \alpha_\nu a_{\nu i} \quad ,$$

which is (17). Furthermore, Theorem II. 1. requires $c_{nn} = O(1)$, i. e. $\bar{b}_{nn}\varepsilon_n a'_{nn} = O(1)$, and this is (18).

The proof of (15) and (16) is similar.

As is easily verified, the $\bar{\varepsilon}_n$ of (15) and the ε_n of (17) are related by

(19) $$\bar{\varepsilon}_n - \bar{\varepsilon}_{n+1} = \Delta\bar{\varepsilon}_n = \varepsilon_n \quad .$$

We show next that (17) and (18) are sufficient in certain cases.

<u>Theorem II. 24.</u> (see [81]) Let A be normal and regular; furthermore assume that $M_K(A)$ holds. Then, if ε_n satisfies (17) the series $\Sigma\, s_n\varepsilon_n$ converges whenever $s_n \in A$.

<u>Proof.</u> We have

$$\sum_{\nu=0}^{n} s_\nu\varepsilon_\nu = \sum_{k=0}^{n} \alpha_k\sigma_k + \sum_{k=n+1}^{\infty} \alpha_k \sum_{\nu=0}^{n} a_{k\nu}s_\nu = \sum_{k=0}^{n} \alpha_k\sigma_k + \sum_{k=n+1}^{\infty} \alpha_k\ O(1) \to \sum_{k=0}^{\infty} \alpha_k\sigma_k$$

because of $M_K(A)$.

<u>Remarks.</u> 1. Condition (18) was not used at all in this proof. This is not surprising when we observe that (18) is a consequence of (17) if $M_K(A)$ is satisfied. In fact, because of $b_{nn} = O(1)$ (B is regular), it follows from (9) and (17) that

$$\varepsilon_n = \sum_{k=n}^{\infty} \alpha_k O(a_{nn}) = o(a_{nn}) = o(\frac{a_{nn}}{b_{nn}}) \quad .$$

2. Theorem II. 24. ensures convergence of $\sum_{0}^{\infty} s_n\varepsilon_n$ and, therefore, also B-summability of this series.

We now turn to a discussion of problem (i).

Theorem II. 25. ([43]) Assumptions on A, B and ε_n as in Theorems II. 23. and II. 24. Then $\Sigma a_n \in A$ implies $\Sigma a_n \bar{\varepsilon}_n \in B$ iff

(20) $$s_n \in A \quad \text{implies} \quad s_n \bar{\varepsilon}_n \in B$$

Proof. We have to show that, under $s_n \in A$,

$$\Sigma a_n \bar{\varepsilon}_n \in B \quad \text{iff} \quad s_n \bar{\varepsilon}_n \in B$$

By partial summation,

$$\sum_{\nu=0}^{n} a_\nu \bar{\varepsilon}_\nu = s_n \bar{\varepsilon}_n + \sum_{\nu=0}^{n-1} s_\nu \varepsilon_\nu ,$$

and $\sum_{0}^{\infty} s_\nu \varepsilon_\nu$ converges by Theorem II. 24. .

Remark. It follows from $\sum_{\nu=0}^{n} a_\nu \bar{\varepsilon}_\nu = s_n \bar{\varepsilon}_{n+1} + \sum_{\nu=0}^{n} s_\nu \varepsilon_\nu$ that $s_n \bar{\varepsilon}_n$ in (20) can be replaced by $s_n \bar{\varepsilon}_{n+1}$.

Our next theorem gives a partial solution of problem (i).

Theorem II. 26. (see [45]) Assumptions as in Theorem II. 25. and let $a_{n\nu} > 0$ for $\nu \leq n$. If $\bar{\varepsilon}_n$ satisfies (15) and (16) then $\Sigma a_n \bar{\varepsilon}_n \in B$ whenever $\Sigma a_n \in A$ if

(21) $$0 < \frac{a_{nn}}{b_{nn}} \downarrow \text{ for } n \uparrow , \quad 0 \leq \frac{b_{n\nu}}{a_{n\nu}} \downarrow \text{ for } n \uparrow (n \geq \nu) , \quad \frac{b_{n\nu}}{a_{n\nu}} \uparrow \text{ for } \nu \uparrow (\nu \leq n) .$$

Proof. It is, by Theorem II. 25. sufficient to show that $(b_{n\nu} \bar{\varepsilon}_\nu) \supseteq (a_{n\nu})$, and this is true (Theorem II. 19.) if

$$\sum_{\nu=0}^{n} \left| \Delta \frac{b_{n\nu} \bar{\varepsilon}_\nu}{a_{n\nu}} \right| = O(1) .$$

But (observe (19))

$$\Delta \frac{b_{n\nu} \bar{\varepsilon}_\nu}{a_{n\nu}} = \varepsilon_\nu \frac{b_{n\nu}}{a_{n\nu}} + \bar{\varepsilon}_{\nu+1} \Delta \frac{b_{n\nu}}{a_{n\nu}}$$

and $(\bar{\varepsilon}_{\nu+1} = \bar{\varepsilon}_{\nu+1} - \bar{\varepsilon}_{n+1} + \bar{\varepsilon}_{n+1})$

$$\sum_{\nu=0}^{n} |\bar{\varepsilon}_{\nu+1}| \, |\Delta \frac{b_{n\nu}}{a_{n\nu}}| \leq |\bar{\varepsilon}_{n+1}| \sum_{\nu=0}^{n} |\Delta \frac{b_{n\nu}}{a_{n\nu}}| + \sum_{\nu=0}^{n-1} |\Delta \frac{b_{n\nu}}{a_{n\nu}}| \sum_{\zeta=\nu+1}^{n} |\Delta \bar{\varepsilon}_{\zeta}|$$

$$= O(1)\, \bar{\varepsilon}_{n+1} \frac{b_{nn}}{a_{nn}} + \sum_{\zeta=1}^{n} |\varepsilon_{\zeta}| \sum_{\nu=0}^{\zeta-1} |\Delta \frac{b_{n\nu}}{a_{n\nu}}| \leq O(1) + \sum_{\zeta=1}^{n} |\varepsilon_{\zeta}| \, |\frac{b_{n\zeta}}{a_{n\zeta}}| \quad .$$

It remains to show that $\quad \sum_{\nu=0}^{n} |\varepsilon_{\nu} \frac{b_{n\nu}}{a_{n\nu}}| = O(1) \quad .$

But

$$\sum_{\nu=0}^{n} |\frac{b_{n\nu}}{a_{n\nu}} \sum_{k=\nu}^{\infty} \alpha_k a_{k\nu}| \leq \sum_{k=0}^{n} |\alpha_k| \sum_{\nu=0}^{k} a_{k\nu} \frac{b_{n\nu}}{a_{n\nu}} + \sum_{k=n+1}^{\infty} |\alpha_k| \sum_{\nu=0}^{n} a_{k\nu} \frac{b_{n\nu}}{a_{n\nu}}$$

and

$$\sum_{\nu=0}^{k} a_{k\nu} \frac{b_{n\nu}}{a_{n\nu}} \leq \sum_{\nu=0}^{k} a_{k\nu} \frac{b_{k\nu}}{a_{k\nu}} = \sum_{\nu=0}^{k} b_{k\nu} = O(1) \qquad (n \geq k) \quad .$$

Furthermore, because of $M_K(A)$, we have for some $j \leq n$

$$\sum_{\nu=0}^{n} a_{k\nu} \frac{b_{n\nu}}{a_{n\nu}} \leq K \sum_{\nu=0}^{j} a_{j\nu} \frac{b_{n\nu}}{a_{n\nu}} \leq K \sum_{\nu=0}^{j} a_{j\nu} \frac{b_{j\nu}}{a_{j\nu}} = O(1)$$

and this proves the theorem.

We mention some special cases of Theorem II. 26. and II. 23. . If $B = A$, then the conditions (21) are satisfied, and it follows in particular that $\Sigma a_n \bar{\varepsilon}_n \in C_\alpha$ whenever $\Sigma a_n \in C_\alpha$ $(0 < \alpha \leq 1)$ iff

$$\bar{\varepsilon}_n = c + \sum_{k=n}^{\infty} \alpha_n \frac{A_{k-n}^{\alpha}}{A_k^{\alpha}} \quad , \quad \Sigma |\alpha_n| < \infty \quad . \tag{22}$$

(This theorem is true for all $\alpha \geq 0$. It was proved for $\alpha = 0, 1, \ldots$ by H. Bohr [7] and G. H. Hardy [22] , and in general by A. F. Andersen [4]; see also L. S. Bosanquet [10].)

Using the notation $\Delta^{\gamma} x_n = \sum_{k=n}^{\infty} A_{k-n}^{-\gamma-1} x_k$ (provided that this series converges) one can show that (22) is equivalent to

$$\Sigma (n+1)^{\alpha} |\Delta^{\alpha+1} \bar{\varepsilon}_n| < \infty \quad , \quad \bar{\varepsilon}_n = O(1) \quad . \tag{23}$$

If $B = I = (\delta_{n\nu})$ then (21) is satisfied if $a_{nn}\downarrow$.

If $A = C_\alpha$, $B = C_\beta$, $0 < \beta \leq \alpha \leq 1$, then it is easily verified that (21) is true, and it follows that $\Sigma a_n \bar{\varepsilon}_n \in C_\beta$ whenever $\Sigma a_n \in C_\alpha$ iff (22) and $\bar{\varepsilon}_n = O(n^{\beta-\alpha})$ holds. (This theorem was proven without restriction on α and β by L. S. Bosanquet [13].)

In connection with these examples we have mentioned that a solution of problem (i) is given by (15) and (16) in certain cases where $M_K(A)$ is not satisfied, and we are thus led to the question whether an extension of Theorem II. 26. exists without the assumption $M_K(A)$. There exists an affirmative answer to this problem (G. E. Peterson [79]). We will not go into the details, and restrict ourselves to the proof of a result for $B = A$ (see [48]).

It will be convenient to denote $\sum_{k=n}^{\infty} \alpha_k \bar{a}_{kn}$, $\Sigma|\alpha_k| < \infty$, by $\varepsilon_n(\bar{A}, \alpha)$. We establish some identities first.

<u>Lemma II. 6.</u> Let A and M_p be normal and regular. Writing $C = AM_p$ we have

$$(24)\qquad \sum_{\nu=0}^{n} c_{n\nu}\varepsilon_{\nu+1}(\bar{C},\alpha)s_\nu = \sum_{\nu=0}^{n} a_{n\nu}\varepsilon_{\nu+1}(\bar{C},\alpha)\frac{1}{P_\nu}\sum_{\mu=0}^{\nu} p_\mu s_\mu - \sum_{\nu=1}^{n} c_{n\nu}(\varepsilon_\nu(\bar{C},\alpha) - \varepsilon_\nu(\bar{A},\alpha))\frac{1}{P_{\nu-1}}\sum_{\mu=0}^{\nu-1} p_\mu s_\mu ,$$

$$(25)\qquad \sum_{\nu=0}^{n} a_\nu \varepsilon_\nu(\bar{C},\alpha) = a_0\varepsilon_0(\bar{C},\alpha) + \sum_{\nu=1}^{n}\varepsilon_\nu(\bar{A},\alpha)\hat{P}_\nu + \varepsilon_{n+1}(\bar{C},\alpha)\Big(s_n - \frac{1}{P_n}\sum_{\nu=0}^{n} p_\nu s_\nu\Big)$$

$$\Big(s_n = a_0 + \dots + a_n ,\quad \hat{P}_n = \frac{p_n}{P_n P_{n-1}}\sum_{\nu=1}^{n} P_{\nu-1}a_\nu ,\quad n \geq 1\Big).$$

If $A \subsetneqq C$, then to every $\alpha = \{\alpha_n\}$, $\Sigma|\alpha_n| < \infty$, exists a sequence $\beta = \{\beta_n\}$, $\Sigma|\beta_n| < \infty$, such that

$$(26)\qquad \varepsilon_n(\bar{C},\alpha) = \varepsilon_n(\bar{A},\beta) .$$

<u>Proof.</u> It follows from $C = AM_p$ that $\bar{C} = \bar{A}\hat{M}_p$, where $\hat{M}_p$ is the form of the method M_p which transforms a series into a series. A short calculation shows that $(\hat{M}_p)_{n\nu} = \frac{p_n}{P_n P_{n-1}} P_{\nu-1}$ $(1 \leq \nu \leq n)$, $(\hat{M}_p)_{no} = \delta_{no}$. A consequence of $\bar{C} = \bar{A}\hat{M}_p$ is the easily verified relation $\bar{c}_{kn}(1 + \frac{P_{n-1}}{p_n}) - \frac{P_{n-1}}{p_n}\bar{c}_{k,n+1} = \bar{a}_{kn}$, and it follows that

$$(27) \qquad \frac{P_{n-1}}{P_n}\left(\varepsilon_{n+1}(\bar{C},\alpha)-\varepsilon_n(\bar{C},\alpha)\right) = \varepsilon_n(\bar{C},\alpha) - \varepsilon_n(\bar{A},\alpha) \quad .$$

Introducing $\sum_{\mu=0}^{\nu} p_\mu s_\mu$ by partial summation into the left-hand side of (24) and using (27) shows that (24) is true. Likewise, (25) follows from partial summation (introducing $\sum_{\mu=1}^{\nu} P_{\mu-1} a_\mu$) and application of (27).

If $A \subseteqq C$, then a regular matrix B exists such that $\bar{C} = B\bar{A}$, and (26) follows from $\varepsilon_n(\bar{C},\alpha) = \alpha\bar{C} = (\alpha B)\bar{A} = \beta\bar{A}$ (see the proof of Theorem II. 10.).

<u>Theorem II. 27.</u> Let A and M_p be normal and regular, and let $C = A\,M_p$. Assume that $s_n \in A$ implies $s_n \in C$ and $t_n \in C$ where $t_0 = 0$, $t_n = s_{n-1}$ $(n \geq 1)$. Furthermore, assume that

$$(28) \qquad \begin{cases} \varepsilon_{n+1}(\bar{A},\alpha)s_n \in A & \text{and} \quad \sum a_n\varepsilon_n(\bar{A},\alpha) \in A \\ \text{whenever } s_n \in A & (s_n = a_0 + \ldots + a_n) \end{cases} \quad .$$

Then (28) is also true when $A, \bar{A}$ is replaced by $C, \bar{C}$ throughout (28).

<u>Proof.</u> Let $s_n \in C$, then $\tau_n = \frac{1}{P_n}\sum_{\nu=0}^{n} p_\nu s_\nu \in A$, and $a_0 + \sum^{\infty} \hat{p}_n \in A$ by the definition of C.

It follows from (26) and (28) that the sequences $\{\varepsilon_{n+1}(\bar{C},\alpha)\,\tau_n\}$ and $\{\varepsilon_{n+1}(\bar{A},\alpha)\,\tau_n\}$ are limitable A. This implies, because of the assumptions on A and C, that the sequence $\{(\varepsilon_n(\bar{C},\alpha) - \varepsilon_n(\bar{A},\alpha))\,\tau_{n-1}\}$ is limitable C. We now conclude from (24) that

$$(29) \qquad \varepsilon_{n+1}(\bar{C},\alpha)\, s_n \in C \qquad\qquad s_n \in C \quad .$$

It follows from (28) that $\sum \varepsilon_n(\bar{A},\alpha)\hat{p}_n \in A$, and hence summable C. Furthermore, $s_n - \tau_n \in C$, and (29) implies $\varepsilon_{n+1}(\bar{C},\alpha)(s_n - \tau_n) \in C$. The identity (25) now shows $\sum a_n \varepsilon_n(C,\alpha) \in C$ whenever $s_n \in C$.

It is the purpose of Theorem II. 27. to derive from a solution of problem (i) for A (in the case A = B) a solution for $C = A\,M_p$, and by repeating this procedure, a solution for $A\,(M_p)^j$, $j = 1, 2, \ldots$.

Observe that Theorem II. 25. and II. 26. (and the remark after Theorem II. 25.) provide us with (28) under the assumption $M_K(A)$.

A matrix A is called <u>translative</u> if $s_n \in A$ implies $t_n \in A$, $t_o = 0$, $t_n = s_{n-1}$ $(n \geq 1)$. If A is translative, and if $A \subseteq C = A\,M_p$, then $s_n \in A$ implies $t_n \in C$.

Every regular Nörlund mean is translative. This follows from the relation

$$\frac{1}{P_n}\sum_{\nu=1}^{n} p_{n-\nu}t_\nu = \frac{P_{n-1}}{P_n}\left(\frac{1}{P_{n-1}}\sum_{\nu=0}^{n-1} p_{n-1-\nu}s_\nu\right) .$$

Combining this result with Theorems II. 22. , II. 23. , II. 25. , II. 26. , and II. 27. we obtain the socalled Bohr-Hardy Theorem:

<u>Theorem II. 28.</u> Let $\alpha \geq 0$, then $\Sigma a_n \varepsilon_n \in C_\alpha$ whenever $\Sigma a_n \in C_\alpha$ iff (22) is satisfied.

3. Tauberian theorems.

We generalize Theorem I. 2. (when restricted to slowly oscillating sequences).

<u>Theorem II. 29.</u> Let $A = (a_{n\nu})$ be normal and regular, and satisfy $M_K(A)$. Moreover, assume that $a_{n\nu} \geq 0$ $(\nu \leq n)$ and that $0 \leq \phi(n,m)$ $(m \leq n = 0,1,2,\dots)$ exists such that $\phi(n,m) \downarrow$ for $m \uparrow$ and

$$\liminf_{n\to\infty} \sum_{\substack{\nu\leq n\\ \phi(n,\nu)\leq\varepsilon}} a_{n\nu} = \Phi(\varepsilon) > 0$$

for every $\varepsilon > 0$. Then it follows from $s_n \in A$ and $s_n - s_m \to 0$ $(m \leq n)$ for $\phi(n,m) \to 0$ $(n \to \infty)$ that $\{s_n\}$ converges.

<u>Proof.</u> Let $s_n \to 0$ (A) and assume that $\limsup s_n \geq 2\alpha > 0$. Then, there exists $\varepsilon > 0$ such that for $n \geq N$ and $\phi(n,m) \leq \varepsilon$ the inequality $|s_m - s_n| \leq \frac{\alpha}{2}$ is true. If $s_{n_i} \geq \alpha$ $(n_i \geq N)$, then $s_m \geq \frac{\alpha}{2}$ for $m \leq n_i$, $\phi(n_i, m) \leq \varepsilon$.

It follows that

$$\sum_{\substack{\nu\leq n_i\\ \phi(n_i,\nu)\leq\varepsilon}} a_{n_i\nu}s_\nu \geq \frac{\alpha}{2}\sum_{\substack{\nu\leq n_i\\ \phi(n_i,\nu)\leq\varepsilon}} a_{n_i\nu} \geq \frac{\alpha}{2}\,\frac{1}{2}\,\Phi(\varepsilon)$$

for all large i .

But, for large i, we have the contradiction

$$\sigma_{n_i} = \sum_{\substack{\nu\le n_i\\ \phi(n_i,\nu)\le\varepsilon}} a_{n_i\nu}s_\nu + \sum_{\substack{\nu\le n_i\\ \phi(n_i,\nu)>\varepsilon}} a_{n_i\nu}s_\nu \ge \frac{\alpha}{2}\frac{1}{2}\Phi(\varepsilon)+o(1)$$

because of $M_K(A)$ (and A is perfect!).

Applications. Let $A = M_p$, $p_n > 0$, be regular. If we define $\phi(n,m) = \frac{P_n - P_m}{P_m} = -1 + \frac{P_m}{P_n}$ $(m \le n)$ then $\phi(n,m)\downarrow$ for $m\uparrow$, and for $0<\varepsilon<1$ let ν_0 be such that $P_n - P_{\nu_0} \le \varepsilon P_{\nu_0}$, $P_n - P_{\nu_0-1} > \varepsilon P_{\nu_0-1}$; it follows that

$$\frac{1}{P_n}\sum_{\phi(n,\nu)\le\varepsilon} p_\nu = \frac{1}{P_n}(P_n - P_{\nu_0-1}) \ge \frac{\varepsilon P_{\nu_0-1}}{P_n} \ge \frac{\varepsilon P_{\nu_0-1}}{(1+\varepsilon)P_{\nu_0}},$$

and $\Phi(\varepsilon) > 0$ if $\liminf \frac{P_{n-1}}{P_n} > 0$.

Let $a_n = O(\frac{p_n}{P_n})$, then for $m<n$ $s_n - s_m = O(1)\frac{P_n-P_m}{P_m} = O(1)\,\phi(n,m)$, and it follows that $a_n = O(\frac{p_n}{P_n})$ is a Tauberian condition for M_p if $\liminf \frac{P_{n-1}}{P_n} > 0$. In particular, for C_1 this is the condition $a_n = O(\frac{1}{n})$.

Problems.

1. Show that the relation $M_p \approx N_p$ is true for a monotone p_n iff $M_p \approx C_1$.
2. Show that under the assumptions of Theorem II.21. we have $A \approx B$ if
$$0 < \frac{a_{nn}}{b_{nn}} \le K \quad \text{for some} \quad K > 0.$$
3. Show that $\sum_n \frac{s_n}{3^n}$ is convergent when $s_n \in E_1$ (see [20] [82]).
4. Show that a regular and positive mean M_p is translative iff
$$\sum_{\nu=1}^{n-1} P_{\nu-1}\left|\frac{p_\nu}{p_{\nu-1}} - \frac{p_{\nu+1}}{p_\nu}\right| = o(P_n) \qquad ([21]).$$
5. Give a generalization of Theorem II.29. to slowly decreasing sequences.
6. Show that for N_p, $p_n = \frac{1}{n+1}$, $a_n = O(n^{-\delta})$ is a Tauberian condition for every $\delta > 0$.

7. The Knopp-Schnee-Hausdorff Theorem

In 1907 K. Knopp [50] proved $H^k \subsetneqq C_k$, and in 1909 W. Schnee [95] proved $C_k \subsetneqq H^k$ $(k=0,1,\ldots)$. In 1921 F. Hausdorff [32] proved $H^k \approx C_k$ $(k>-1)$. In this section we will prove the last result for $k \geq 0$.

We recall the following facts.

1°. If A, B and C are normal, and if $A \approx B$ (i.e., AB^{-1} and BA^{-1} are regular), then $AC \approx BC$ (since $AC(BC)^{-1} = ACC^{-1}B^{-1} = AB^{-1}$ is regular, and similarly for $BC(AC)^{-1}$; see the proof of Theorem II.22.).

2°. Any two Hausdorff methods commute.

The equivalence $C_k \approx H^k$ $(k \geq 0)$ is a consequence of the following two relations:

(i) $\quad C_{k+1} \approx C_k C_1 \qquad (k \geq 0)$

(ii) $\quad C_k \approx H^k \qquad (0 \leq k \leq 1)$.

In fact, if $p < k \leq p+1$ $(p=0,1,2,\ldots)$ then

$$C_k \approx C_{k-1}C_1 \approx C_{k-2}C_1^2 \approx \ldots \approx C_{k-p}(C_1)^p \approx H^{k-p}H^p = H^k .$$

It remains to show that (i) and (ii) are true. Relation (i) was proved in Theorem II.22., and we turn to (ii).

Lemma II.7. Assume that $0 \leq F(x) \uparrow$ $(0 \leq x \leq 1)$ is absolutely continuous, $F(1) < \infty$. Furthermore, let $a(x), b(x) \in L(0,1)$, $a(x), b(x) > 0$ $(0<x<1)$ and $\frac{a(x)}{b(x)} \downarrow$ as $x \uparrow$ $(0<x<1)$. Then

$$\frac{\int_0^1 F(x)a(x)dx}{\int_0^1 a(x)dx} \leq \frac{\int_0^1 F(x)b(x)dx}{\int_0^1 b(x)dx} . \tag{30}$$

Proof. We may assume that $F(0) = 0$ - otherwise replace $F(x)$ by $F(x) - F(0)$ which does not affect (30). Let

$$A(x) = \int_x^1 a(t)\,dt , \qquad B(x) = \int_x^1 b(t)\,dt .$$

By partial integration, $\int_0^1 F(x)\,a(x)\,dx = \int_0^1 F'(x)\,A(x)\,dx$, and similarly for b. But

$$\frac{\int_0^1 F(x)a(x)dx}{\int_0^1 F(x)b(x)dx} = \frac{\int_0^1 F'(x)B(x)\,\frac{A(x)}{B(x)}\,dx}{\int_0^1 F'(x)B(x)\,dx} \le \sup_{0\le x\le 1} \frac{A(x)}{B(x)}$$

Next,

$$\frac{d}{dx}\,\frac{A(x)}{B(x)} = \frac{b(x)}{B^2(x)} \int_x^1 b(t)\left(\frac{a(t)}{b(t)} - \frac{a(x)}{b(x)}\right)dt \le 0$$

(almost everywhere), and it follows that $\sup \frac{A(x)}{B(x)} = \frac{A(0)}{B(0)}$. This proves the lemma.

Remark. It follows from the proof of Lemma II. 7. that the condition $F(1) < \infty$ can be omitted if $F(x)A(x) \to 0$, $F(x)\,B(x) \to 0$ as $x \to 1$.

Lemma II. 8. $\dfrac{\log \frac{1}{x}}{1-x} \downarrow$ for $x \uparrow$ $(0 < x < 1)$.

Proof. By differentiation and $x \le e^{x-1}$ for $x \ge 1$

We will prove now an inequality for the terms of the matrices C_k and H^k.

Let $0 < k < 1$, $F(x) = \left(\dfrac{\log \frac{1}{x}}{1-x}\right)^{k-1}$, $a(x) = (1-x)^{k-1} x^{\nu}\,(1-x)^{n-\nu}$, $b(x) = (1-x)^{k-1} x^{\nu+1} (1-x)^{n-\nu-1}$ $(0 \le \nu < n)$.

Here $0 \le F(x) \uparrow$ and $\frac{a(x)}{b(x)} = \frac{1}{x} - 1 \downarrow$ for $x \uparrow$, and (30) turns out to be (see page 22)

(31) $$\theta_{n\nu} = \frac{H^k_{n\nu}}{(C_k)_{n\nu}} \le \frac{H^k_{n,\nu+1}}{(C_k)_{n,\nu+1}}.$$

It follows from Theorem II.21. that $H^k \subsetneqq C_k$, and Theorem II. 19. yields $C_k \subsetneqq H^k$ (note that $\theta_{nn} = O(1)$). We thus have

Theorem II. 30. (Knopp-Schnee-Hausdorff) $C_k \approx H^k$ for $k \ge 0$

It follows from Theorems II. 18. and II. 30. that $M_L(H^{\alpha})$, $0 \le \alpha \le 1$, holds for some $L > 0$, and the question arises whether $M_1^*(H^{\alpha})$ is true. We are now in a position to prove $M_1^*(H^{\alpha})$; more generally we will discuss conditions which guarantee $M_1^*(H)$ for a Hausdorff method H.

Let $H = \Delta D \Delta$ be a regular Hausdorff method with $d_n = \int_0^1 \phi(t)\, t^n\, dt$, $0 \leq \phi(t) \in L(0,1)$, and assume that $a_{n\nu} > 0$ $(\nu \leq n)$.

The inequality $\frac{a_{n+1,\nu}}{a_{n\nu}} \geq \frac{a_{n+1,\nu+1}}{a_{n,\nu+1}}$ $(\nu < n)$ of Lemma II. 5. is in this case $(\nu < n)$

$$(32) \qquad \frac{n-\nu}{n+1-\nu} \frac{\int_0^1 \phi(t) t^{\nu}(1-t)^{n+1-\nu} dt}{\int_0^1 \phi(t) t^{\nu}(1-t)^{n-\nu} dt} \geq \frac{\int_0^1 \phi(t) t^{\nu+1}(1-t)^{n-\nu} dt}{\int_0^1 \phi(t) t^{\nu+1}(1-t)^{n-\nu-1} dt} .$$

Partial integration yields

$$\int_0^1 \phi(t) t^{\nu}(1-t)^{n+1-\nu} dt = (n+1-\nu) \int_0^1 (1-t)^{n-\nu} \int_0^t \phi(\tau)\tau^{\nu} d\tau\, dt ,$$

and a similar formula holds for the other integral on the left-hand side. Using these formulas the inequality (32) is of the form (30) with

$$a(x) = (1-x)^{n-\nu} \int_0^x \phi(\tau)\tau^{\nu} d\tau , \quad b(x) = (1-x)^{n-\nu-1} \int_0^x \phi(\tau)\tau^{\nu} d\tau , \quad F(x) = \frac{\phi(x) x^{\nu+1}}{\int_0^x \phi(\tau)\tau^{\nu} d\tau} ,$$

and $M_1^*(A)$ holds if $F(x) \uparrow$ and $\phi(x)(1-x) \to 0$ for $x \uparrow 1$ (we omit the simple proof that $F(x)A(x) \to 0$, $F(x)B(x) \to 0$ if $\phi(x)(1-x) \to 0$ for $x \to 1$). If $\phi(x)\, x \to 0$ for $x \to 0$, and if $\phi(x)$ is absolutely continuous, then it follows from

$$\left(\int_0^x \phi(\tau)\tau^{\nu} d\tau\right)^2 \frac{d}{dx} F(x) = \phi(x) x^{\nu} \int_0^x \phi(\tau)\tau^{\nu} \left\{ \frac{(\phi(x) x^{\nu+1})'}{\phi(x) x^{\nu}} - \frac{(\phi(\tau)\tau^{\nu+1})'}{\phi(\tau)\tau^{\nu}} \right\} d\tau \geq 0$$

(almost everywhere) that $F(x) \uparrow$ is a consequence of

$$\frac{(x^{\nu+1}\phi)'}{x^{\nu}\phi} = (\nu+1) + x \frac{\phi'(x)}{\phi(x)} \uparrow \quad \text{as} \quad x \uparrow .$$

For H^k $(0 < k \leq 1)$ we have $\Gamma(k)\, \phi(x) = (\log \frac{1}{x})^{k-1}$ and

$$x \frac{\phi'(x)}{\phi(x)} = \frac{1-k}{\log \frac{1}{x}} \uparrow \quad \text{as} \quad x \uparrow .$$

This implies $M_1^*(H^k)$ for $0 < k \leq 1$.

A final remark about (i) for $k = 1$. The method $H^2 C_2^{-1} = C_1 M_p^{-1}$, $p_n = (n+1)$, is equivalent to convergence. A short calculation shows that $C_1 M_p^{-1} = \frac{1}{2}(C_1 + C_0)$, i. e., the method $\sigma_n = \frac{1}{2}(\frac{s_0 + \ldots + s_n}{n+1} + s_n)$

is equivalent to convergence. This was proven in 1906 by J. Mercer [74], and every theorem which states that a method is equivalent to convergence is called a "Mercerian" Theorem (see also problem 4 on page 36).

Problems.

1. Show that $H^k = \left(\prod_{\nu=2}^{k} \left(\frac{1}{\nu} I + \frac{\nu-1}{\nu} C_1\right)\right) C_k$, $I = (\delta_{n\nu})$, and all factors in this product are equivalent to convergence (I. Schur [96]).
2. Show that $(H^k C_k^{-1})_{n\nu} \leq 0$, $(C_k H^{-k})_{n\nu} \geq 0$ for $0 \leq \nu < n$, $0 < k \leq 1$.
3. Show that $s_n \to \infty$ (C_k) implies $s_n \to \infty$ (H^k) for $k = 1, 2, \ldots$ (see I. Schur [96], K. Knopp [55]).
4. Show that $s_n \to \infty$ (H^k) implies $s_n \to \infty$ (C_k) when $0 < k \leq 1$ (S. K. Basu [6]).

8. Functions of a matrix

In section 9 of this chapter we will discuss some properties of powers A^k (k not necessarily an integer) of a normal and regular matrix A . These powers are naturally defined when k is an integer, and in the present section we will also define powers A^k when k is not an integer. More generally, we will define functions f(A) of a matrix A (see [47]).

In order to motivate the following formulas we write the powers A^2 and A^3 of a matrix A in a certain form. Our notation is $A = (a_{n\nu})$, $A^2 = (A^2_{n\nu})$, $A^3 = (A^3_{n\nu})$, $a_n = a_{nn}$, $d^a_{n\nu_1 \cdots \nu_k \nu} = a_{n\nu_1} a_{\nu_1 \nu_2} \cdots a_{\nu_k \nu}$ for $n \neq \nu_1$, $\nu_1 \neq \nu_2, \ldots, \nu_k \neq \nu$, and

$d^a_{n\nu_1\cdots\nu_k\nu} = 0$ otherwise, and we have (formally)

$$A^2_{n\nu} = \sum_{\nu_1} a_{n\nu_1} a_{\nu_1\nu} = a_n a_{n\nu} + a_{n\nu} a_\nu (1-\delta_{n\nu}) + \sum_{\nu_1} d^a_{n\nu_1\nu} = \delta_{n\nu} a_n^2 + (a_n + a_\nu) d^a_{n\nu} + \sum_{\nu_1} d^a_{n\nu_1\nu} ,$$

and similarly from a short calculation

$$A^3_{n\nu} = \sum_{\nu_1} a_{n\nu_1} A^2_{\nu_1\nu} = \delta_{n\nu} a_n^3 + (a_n^2 + a_n a_\nu + a_\nu^2) d^a_{n\nu} + \sum_{\nu_1} (a_n + a_{\nu_1} + a_\nu) d^a_{n\nu_1\nu} + \sum_{\nu_1\nu_2} d^a_{n\nu_1\nu_2\nu} .$$

Using the notation of a divided difference $[x_0,\dots,x_n]_f$ (i.e. $[x_0]_f = f(x_0)$, $[x_0, x_1]_f (x_0 - x_1) = f(x_0) - f(x_1)$, $[x_0,\dots x_n]_f (x_0 - x_n) = [x_0,\dots,x_{n-1}]_f - [x_1,\dots,x_n]_f$ when the x_i's are all different; another definition of $[x_0,\dots,x_n]_f$ is

$$[x_0,\dots,x_n]_f = \sum_{i=0}^{n} \frac{f(x_i)}{\phi'(x_i)} , \quad \phi(x) = (x-x_0)\dots(x-x_n) \;;$$ when the x_i' s are not all different,

then $[x_0,\dots,x_n]_f$ is obtained by a limit process) we have

$$A^2_{n\nu} = \delta_{n\nu}[a_n]_{x^2} + d^a_{n\nu} [a_n a_\nu]_{x^2} + \sum_{\nu_1} d^a_{n\nu_1\nu} [a_n a_{\nu_1} a_\nu]_{x^2} ,$$

$$A^3_{n\nu} = \delta_{n\nu}[a_n]_{x^3} + d^a_{n\nu} [a_n a_\nu]_{x^3} + \sum_{\nu_1} d^a_{n\nu_1\nu} [a_n a_{\nu_1} a_\nu]_{x^3} + \sum_{\nu_1\nu_2} d^a_{n\nu_1\nu_2\nu} [a_n a_{\nu_1} a_{\nu_2} a_\nu]_{x^3} .$$

From these formulas we expect that for $k = 0, 1, 2, \dots$

$$(33) \qquad A^k_n = \delta_{n\nu} [a_n]_{x^k} + d^a_{n\nu} [a_n a_\nu]_{x^k} + \sum_{k=1}^{\infty} \sum_{\nu_1\dots\nu_k} d^a_{n\nu_1\nu_2\dots\nu_k\nu} [a_n a_{\nu_1} \dots a_\nu]_{x^k} ,$$

and it seems natural, to <u>define</u> $f(A)|_{n\nu}$ (i.e. the element in position n,ν of the matrix $f(A)$ for a given f) by

$$(34) \qquad f(A)|_{n\nu} = \delta_{n\nu} [a_n]_f + d^a_{n\nu} [a_n a_\nu]_f + \sum_{k=1}^{\infty} \sum_{\nu_1\dots\nu_k} d^a_{n\nu_1\dots\nu_k\nu} [a_n a_{\nu_1} \dots a_\nu]_f .$$

A short induction-type proof shows that (33) is correct - this proof is left to the reader. In connection with definition (34) we assume that $f(z)$ is defined and regular for $z = a_0, a_1, \dots$ (that implies that $f(z)$ is defined in a neighborhood of every a_i - we require regularity in order to be certain that $[a_0,\dots,a_n]_f$ exists also in the case where the a_i' s are not different) , and if the sums occurring in (34) are absolutely convergent. If A is triangular, then these sums are finite since $\nu < \nu_k < \nu_{k-1} < \dots < \nu_1 < n$ $\quad (k \leq n-\nu-1)$.

If A is triangular, then it follows from (33) and (34) that

(35) $$f(A) + g(A) = h(A) \quad \text{for} \quad h(x) = f(x) + g(x) \quad ,$$

(36) $$cf(A) = h(A) \quad \text{for} \quad h(x) = cf(x)$$

(37) $$p(A) = \sum_{k=0}^{n} p_k A^k \quad \text{for} \quad p(x) = \sum_{k=0}^{n} p_k x^k \quad .$$

Furthermore (if A is triangular)

(38) $$f(A)g(A) = h(A) \quad \text{if} \quad h(x) = f(x)g(x) \quad ,$$

(39) $$f(g(A)) = h(A) \quad \text{if} \quad h(x) = f(g(x)) \quad .$$

A proof of (38) (and similarly of (39)) follows from the observation that (38) is true for polynomials f, g , and that $F(A)|_{n\nu} = G(A)|_{n\nu}$ if $F^{(i)}(a_\mu) = G^{(i)}(a_\mu)$ for $\nu \leq \mu \leq n$ and $i = 0, 1, \ldots, n-\nu$. If f_0, g_0 are polynomials with $f_0^{(i)}(a_\mu) = f^{(i)}(a_\mu)$, $g_0^{(i)}(a_\mu) = g^{(i)}(a_\mu)$, $h_0(x) = f_0(x)\, g_0(x)$ $(\nu \leq \mu \leq n,\ i=0,1,\ldots,n-\nu)$, then

$$f(A)g(A)|_{n\nu} = \sum_{\mu=\nu}^{n} f(A)|_{n\mu} g(A)|_{\mu\nu} = \sum_{\mu=\nu}^{n} f_0(A)|_{n\mu} g_0(A)|_{\mu\nu} = h_0(A)|_{n\nu} = h(A)|_{n\nu}$$

since $h_0^{(i)}(a_\mu) = (f_0(a_\mu)\, g_0(a_\mu))^{(i)} = (f(a_\mu)\, g(a_\mu))^{(i)} = h^{(i)}(a_\mu)$ for $\nu \leq \mu \leq n$, $i=0,1,\ldots,n-\nu$.

We mention three consequences of these formulas.

1. If A is triangular and $a_n > 0$, then $A^\alpha = f(A)$, $f(x) = x^\alpha$, exists for every real α and we have

$$A^\alpha A^\beta = A^{\alpha+\beta} \quad , \quad (A^\alpha)^\beta = A^{\alpha\beta} \quad , \quad A^\alpha = e^{\alpha \log A} \quad .$$

2. A short proof shows that the formula

$$[x_0^{-1}, x_1^{-1}, \ldots, x_n^{-1}]_{g(\frac{1}{x})} = (-1)^n x_0 x_1, \ldots, x_n \ [x_0, \ldots, x_n]_{x^{n-1} g(x)} \qquad (x_i \neq 0)$$

is true. If A is normal, then it follows from (34) and $f(A) = F(A')$, $F(x) = f(\frac{1}{x})$ $(A' = A^{-1})$ that

$$(40) \qquad f(A)|_{n\nu} = a_n \delta_{n\nu} [a_n]_{fx^{-1}} - d^{a'}_{n\nu} a_n a_\nu [a_n a_\nu]_f + \sum_{k=1}^{\infty} (-1)^{k+1} \sum_{\nu_1,\dots,\nu_k} d^{a'}_{n\nu_1 \dots \nu_k \nu} a_n \dots a_\nu \times$$
$$\times [a_n, \dots, a_\nu]_{fx^k} \; .$$

3. If A is triangular, then we write $A \dot{\geqq} 0$ if $a_{n\nu} \geqq 0$ for $\nu < n$, $A \dot{\leqq} 0$ if $a_{n\nu} \leqq 0$ for $\nu < n$.

<u>Theorem II. 31.</u> Let A be normal, $A \dot{\leqq} 0$, $a_n > 0$.
Then $f(A) \dot{\geqq} 0$ if $f(x)$ is <u>completely monotone</u> (i.e. $(-1)^k f^{(k)}(x) \geqq 0$) for $x > 0$, and $f(A) \dot{\leqq} 0$ if $f'(x)$ is <u>completely monotone</u> for $x > 0$.

The <u>proof</u> follows immediately from (34) if we observe that $[a_n a_{\nu_1}, \dots, a_{\nu_k} a_\nu]_f = f^{(k+1)}(\theta) / (k+1)!$ for some $\theta > 0$ (note that $f(z)$ is regular for $\mathrm{Re}\, z > 0$ by a theorem of Bernstein, see [106]).

<u>Remark.</u> The function $f(x) = x^\alpha$ is completely monotone for $\alpha \leqq 0$, and f' is completely monotone for $0 < \alpha \leqq 1$. Let A be normal, $a_n > 0$, $A' \dot{\leqq} 0$, then $A^\alpha \dot{\geqq} 0$ for $\alpha \geqq 0$ and $A^{-\beta} \dot{\leqq} 0$ for $0 \leqq \beta \leqq 1$. (For the connection between the conditions $A' \dot{\leqq} 0$ and $M_1^*(A)$ see Theorems II. 16. and II.17. .)

<u>Problems.</u>

1. Let A be normal with $a_n > 0$, and assume that normal matrices $A(\alpha)$ $(-\infty < \alpha < \infty)$ exist such that $A(\alpha)|_{nn} > 0$, $A(1) = A$, $A(\alpha)\,A(\beta) = A(\alpha + \beta)$. Show that $A(\alpha) = A^\alpha$, where A^α is the matrix given by (34) for $f(x) = x^\alpha$.
2. Give an example of a normal matrix $A \dot{\geqq} 0$, $a_n > 0$ where $A^{1/2} \dot{\geqq} 0$ is false.

9. A generalization of the Knopp-Schnee-Hausdorff Theorem

The Knopp-Schnee-Hausdorff Theorem states the equivalence $(C_1)^\alpha \approx C_\alpha$, where C_α is obtained from C_1 by a certain iteration-process (see chapter I. 2.). One might ask whether a similar relation holds when C_1 is replaced by some other matrix A , and when an iteration of the " Cesàro type" has been defined for A (note that we have defined A^α in the previous section). Before we go into any details we will prove some auxiliary results.

Lemma II. 8. Let A be normal, $A \geq 0$, $a_n > 0$, $A^\alpha = (a_{n\nu}^{(\alpha)})$.

a) If $\sum_{\nu=0}^{n} a_{n\nu} = 1$, then $\sum_{\nu=0}^{n} a_{n\nu}^{(\alpha)} = 1$ for every (real) α .

b) If $a_{n\nu} \to 0$ ($n \to \infty$, ν fixed) and $A^\alpha \geq 0$ for $0 \leq \alpha \leq 1$, then $a_{n\nu}^{(\alpha)} \to 0$ ($n \to \infty$, ν fixed) for $0 \leq \alpha \leq 1$.

Proof. a) It follows from $A^\alpha A = A A^\alpha$ that

$$\sum_{\nu=0}^{n} \sum_{\mu=\nu}^{n} a_{n\mu}^{(\alpha)} a_{\mu\nu} = \sum_{\nu=0}^{n} \sum_{\mu=\nu}^{n} a_{n\mu} a_{\mu\nu}^{(\alpha)} ,$$

and this implies

$$(41) \qquad \sum_{\mu=0}^{n} a_{n\mu}^{(\alpha)} = \sum_{\mu=0}^{n} a_{n\mu} \sum_{\nu=0}^{\mu} a_{\mu\nu}^{(\alpha)} .$$

If $\sum_{\nu=0}^{p} a_{p\nu}^{(\alpha)} = 1$ for $p=0,1,\dots,n-1$ (we have $a_{00}^{(\alpha)} = 1$) , then it follows from (41) that $(1-a_n) \sum_{\mu=0}^{n} a_{n\mu}^{(\alpha)} = \sum_{\mu=0}^{n-1} a_{n\mu} = 1 - a_n$. If $a_n \neq 1$, then $\sum_{\mu=0}^{n} a_{n\mu}^{(\alpha)} = 1$. If $a_n = 1$, then $a_{n\mu} = 0$ for $\mu < n$, and it follows from (33) that $\sum_{\mu=0}^{n} a_{n\mu}^{(\alpha)} = a_{nn}^{(\alpha)} = 1$.

b) We have $a_{n\nu} = \sum_{\mu=\nu}^{n} a_{n\mu}^{(\alpha)} a_{\mu\nu}^{(1-\alpha)} \geq a_{n\nu}^{(\alpha)} a_{\nu\nu}^{(1-\alpha)}$ which implies the statement in b) .

Lemma II. 9. Let $0 < x_0 \leq x_1 \leq \dots\dots$, $0 < \alpha < 1$, $f(x) = x^{n+\alpha-1}$, then

$$\frac{[x_0,\dots,x_n]_f}{\binom{n+\alpha-1}{n}} \downarrow \quad \text{for } n \uparrow$$

Proof. A short calculation shows that

$$[x_0,\dots,x_n]_f = [x_0,\dots,x_{n+1}]_{xf} - x_{n+1} [x_0,\dots,x_{n+1}]_f$$

Using this result we have

$$A\,[x_0,\dots,x_n]_f + B\,[x_0,\dots,x_{n+1}]_{xf} = A\,[x_0,\dots,x_{n+1}]_{xf} - Ax_{n+1}[x_0,\dots,x_{n+1}]_f + B[x_0,\dots,x_{n+1}]_{xf}$$

$$= [x_0,\dots,x_{n+1}]_{h(x)} = h^{(n+1)}(\theta)\,/\,(n+1)! \qquad (x_0 \leq \theta \leq x_{n+1})$$

where $h(x) = A(x-x_{n+1})\,f(x) + Bxf(x)$.

If $A = \binom{n+\alpha}{n+1}$, $B = -\binom{n+\alpha-1}{n}$, then

$$h^{(n+1)}(\theta) = \binom{n+\alpha}{n+1}\,\theta^{\alpha-1}\,(\,n+\alpha-1\ \dots\ \alpha-1\,)\,(1-\frac{x_{n+1}}{\theta}) \geq 0 \quad ,$$

and the statement of the lemma follows.

Starting from a normal and regular matrix A with $a_n > 0$ we write

$$B = (\frac{a_{n\nu}}{a_n}) \quad , \quad B^{\alpha} = (b_{n\nu}^{(\alpha)}) \quad , \quad B_n^{(\alpha)} = \sum_{\nu=0}^{n} b_{n\nu}^{(\alpha)} \qquad A_{\alpha} = (b_{n\nu}^{(\alpha)}\,/\,B_n^{(\alpha)}) \quad .$$

The matrix A_α might be called a "Cesàro iteration" of A : If $A = C_1$, then $A_\alpha = C_\alpha$ (the idea is to normalize A such that the diagonal terms are 1 , then form the α-th power of this normalized matrix, and finally divide it by its row-sums in order to make the row-sums of the resulting matrix 1). - Of course, this is not the only way to define a "Cesàro iteration" .

In view of the Knopp-Schnee-Hausdorff Theorem it seems reasonable to ask whether $A^{\alpha} \approx A_{\alpha}$. We will investigate this question only in the case where A is a mean M_p , and for $0 < \alpha \leq 1$ (the restriction on α means that we essentially discuss a generalization of the Hausdorff contribution to the Knopp-Schnee-Haudorff Theorem).

Let $C = (c_{n\nu})$, $c_{n\nu} = p_\nu / q_n$ for $\nu \leq n$, $c_{n\nu} = 0$ for $\nu > n$ $(p_\nu > 0\,,\ q_\nu > 0)$, then $c'_{nn} = q_n/p_n$, $c'_{n,n-1} = -q_{n-1}/p_n$ $(n \geq 1)$, $c'_{n\nu} = 0$ otherwise, and it follows that $C' \gtrless 0$. By the remark after Theorem II. 31. we have $C^{\alpha} \gtrless 0$ $(\alpha \geq 0)$, $C^{-\alpha} \gtrless 0$ $(0 \leq \alpha \leq 1)$, and from (40) we obtain

$$c_{n\nu}^{\alpha} = (-1)^{n-\nu}(-1)^{n-\nu}\,\frac{q_{n-1}}{p_n}\,\frac{q_{n-2}}{p_{n-1}}\cdots\frac{q_\nu}{p_{\nu+1}}\,\frac{p_n}{q_n}\cdots\frac{p_\nu}{q_\nu}\left[\frac{p_n}{q_n}\cdots\frac{p_\nu}{q_\nu}\right]_{x^{\alpha+n-\nu-1}}$$

$$= \frac{q_\nu}{q_n}\left[\frac{p_n}{q_n},\dots,\frac{p_\nu}{q_\nu}\right]_{x^{n-\nu+\alpha-1}}$$

For $q_n = P_n$, i.e. for $C = M_p$, this leads to

$$M_p^\alpha\Big|_{n\nu} = \frac{p_\nu}{P_n}\left[\frac{p_n}{P_n},\ldots,\frac{p_\nu}{P_\nu}\right]_x x^{n-\nu+\alpha-1} ,$$

and for $q_n = p_n$, i.e. $C = B = ((M_p)_{n\nu} / (M_p)_{nn})$ we have

$$b_{n\nu}^{(\alpha)} = \frac{p_\nu}{p_n}\binom{n-\nu+\alpha-1}{n-\nu} .$$

Theorem II. 32. ([47]) Let M_p be regular and assume that $p_n > 0$, $\frac{p_n}{P_n}\downarrow$. Then $M_p^\alpha \subsetneqq (M_p)_\alpha$ for $0 \le \alpha \le 1$, and $M_p^\alpha \approx (M_p)_\alpha$ iff $\sum_{\nu=0}^{n}\frac{p_\nu}{p_n}\binom{n-\nu+\alpha-1}{n-\nu} = O((\frac{P_n}{p_n})^\alpha)$.

Proof. It follows from Lemma II. 8. that M_p^α is regular; furthermore, we have $(M_p^\alpha)' \gtrless 0$ for $0 \le \alpha \le 1$. Also $(M_p)'_\alpha \gtrless 0$ for $0 \le \alpha \le 1$, and $(M_p)_\alpha$ is regular if $b_{n\nu}^{(\alpha)} = o(B_n^{(\alpha)})$. This is true because of

$$b_{n\nu}^{(\alpha)} / B_n^{(\alpha)} \le \frac{p_\nu\binom{n-\nu+\alpha-1}{n-\nu}}{\sum\limits_{\nu\le\frac{n}{2}} p_\nu\binom{n-\nu+\alpha-1}{n-\nu}} = O(P_m^{-1}) = o(1) \qquad (m \le n/2) .$$

From Lemma II.9. we obtain

$$\frac{M_p^\alpha\big|_{n\nu}}{(M_p)_\alpha\big|_{n\nu}} = B_n^{(\alpha)}\frac{p_n}{P_n}\frac{1}{\binom{n-\nu+\alpha-1}{n-\nu}}\left[\frac{p_n}{P_n},\ldots,\frac{p_\nu}{P_\nu}\right]_x x^{n-\nu+\alpha-1}\uparrow \qquad \text{for } \nu\uparrow ,$$

which implies $M_p^\alpha \subsetneqq (M_p)_\alpha$ by Theorem II. 21. or Theorem II. 20. . If $M_p^\alpha\big|_{nn} / (M_p)_\alpha\big|_{nn} = O(1)$, i.e. if $(\frac{p_n}{P_n})^\alpha B_n^{(\alpha)} = O(1)$, then it follows from Theorem II. 19. that also $M_p^\alpha \supsetneqq (M_p)_\alpha$ is true. Finally, we note that $A \subsetneqq B$ (A, B normal and regular) implies $b_n = O(a_n)$ because of $(BA^{-1})_{nn} = b_n a_n^{-1}$, and this shows that $B_n^{(\alpha)} = O((P_n/p_n)^\alpha)$ is necessary for $M_p^\alpha \approx (M_p)_\alpha$.

The equivalence $M_p^\alpha \approx (M_p)_\alpha$ holds under the assumptions of Theorem II. 32. if $p_n\uparrow$. This follows from

$$\sum_{\nu\le n}\frac{p_\nu}{p_n}\binom{n-\nu+\alpha-1}{n-\nu} = \Big(\sum_{\nu\le\eta_n} + \sum_{\eta_n\le\nu\le n}\Big)\frac{p_\nu}{p_n}\binom{n-\nu+\alpha-1}{n-\nu} = O(\frac{P_n}{p_n}(n-\eta_n)^{\alpha-1}) + O((n-\eta_n)^\alpha)$$

$$\text{for } n-\eta_n \sim \frac{P_n}{p_n} .$$

On the other hand, there are cases where $M_p^\alpha \approx (M_p)_\alpha$ is false. As an example we mention $p_n = \frac{1}{n+1}$; it is easily verified that in this case $B_n^{(\alpha)} \neq O((p_n/P_n)^\alpha)$ $(0 < \alpha < 1)$.

For an extension of Theorem II. 32. to a more general class of matrices see [47] .

Problems.

1. Show that Lemma II. 8. a) is true without the assumption $A \overset{\bullet}{\geq} 0$.

2. Show that $M_p^{\alpha} (M_p)_{\alpha}^{-1} \overset{\bullet}{\leq} 0$, $(M_p)_{\alpha} M_p^{-\alpha} \overset{\bullet}{\geq} 0$ under the assumptions of Theorem II. 32. .

10. Summability functions

Here we will discuss some results of G. G. Lorentz ([65] , [66] , [67] , [68] , [69]) which show that in many cases Tauberian conditions are best possible.

For a sequence $\{s_n\}$ we introduce a " counting function" by

$$A(n,s) = \sum_{\substack{\nu \leq n \\ s_\nu \neq 0}} 1 \quad .$$

The key to the following results is

Lemma II. 10. ([66]) If $0 < F(n) \uparrow \infty$ $(n=0,1,2,\ldots)$, $\Sigma \frac{1}{F(n)} = \infty$, then there exists a bounded and divergent sequence $s = \{s_\nu\}$ such that

$$A(n,s) \leq F(n) \quad , \quad s_n - s_{n-1} = O\left(\frac{1}{F(n)}\right) \quad .$$

Proof. Given m with $F(m) \geq 9$ we define an integer $n = \phi(m) > m$ as follows.

I. If

$$\psi(m) = \left\{\sup k \mid \nu+1-m \leq \frac{1}{2}F(\nu) \quad \text{for} \quad m < \nu \leq k\right\} < \infty \quad ,$$

then

$$\phi(m) = \psi(m)$$

($\psi(n)$ exists since for $k = m+1$, $\nu = k$: $\nu+1-m = 2 \leq \frac{1}{2}F(m+1) \geq \frac{9}{2}$).

II. If $\psi(m) = \infty$, then select $\phi(m)$ such that

$$\sum_{m \leq \nu \leq \phi(m)} \frac{1}{F(\nu)} \geq \frac{1}{3} \quad . \tag{42}$$

Inequality (42) also holds in case I. since

$$\psi(m) + 2 - m \geq \frac{1}{2}F(\psi(m)+1) \quad ;$$

it follows that

$$\sum_{m \leq \nu \leq \phi(m)} \frac{1}{F(\nu)} \geq \frac{\psi(m)+2-m-1}{F(\psi(m)+1)} \geq \frac{1}{2} - \frac{1}{9} > \frac{1}{3} \quad .$$

We construct a sequence $m_1 < n_1 < m_2 < n_2 < \dots$ ($n_i = \phi(m_i)$, $F(m_1) \geq 9$) such that

$$\sum_{m_i \leq \nu \leq n_i} \frac{1}{F(\nu)} \geq \frac{1}{3}$$

and that

$$A(n,t) \leq F(n) \quad \text{for} \quad t_n = \begin{cases} 1 , & m_i \leq n \leq n_i \\ 0 & \text{otherwise} \end{cases} \quad .$$

Given $m_1,\dots,m_i$ and $n_j = \phi(m_j)$, $j = 1,\dots,i$, let m_{i+1} be such that

$$h_i = \sum_{j=1}^{i} (n_j - m_j + 1) \leq \frac{1}{2}F(m_{i+1}) \quad , \quad m_{i+1} > n_i \quad .$$

We have for every m_i and $m_i \leq n \leq n_i$

$$A(n,t) = n+1-m_1 \leq \frac{1}{2}F(n) \leq F(n) \quad ,$$

and for $m_{i+1} \leq n \leq n_{i+1}$ we have

$$A(n,t) = h_i + (n+1-m_{i+1}) \leq \frac{1}{2}F(m_{i+1}) + \frac{1}{2}F(n) \leq F(n) \quad .$$

Finally, we select a sequence ℓ_i , $m_i < \ell_i < n_i$ such that

$$\sum_{m_i \leq n \leq \ell_i} \frac{1}{F(n)} \geq \frac{1}{9} \quad , \quad \sum_{\ell_i < n \leq n_i} \frac{1}{F(n)} \geq \frac{1}{9}$$

(this is possible because of $F(n) \geq 9$).

Choose numbers $0 < \alpha_i < 1$, $-1 < \beta_i < 0$ such that

$$\sum_{m_i \leq n \leq \ell_i} \frac{\alpha_i}{F(n)} = \frac{1}{9} \quad , \quad \sum_{\ell_i < n \leq n_i} \frac{\beta_i}{F(n)} = -\frac{1}{9} \quad .$$

Writing $a_n = \frac{\alpha_i}{F(n)}$ for $m_i \leq n \leq \ell_i$, $a_n = \frac{\beta_i}{F(n)}$ for $\ell_i < n \leq n_i$, $a_n = 0$ for $n_i < n < m_{i+1}$, we have $a_n = O(\frac{1}{F(n)})$, $|s_n| \leq \frac{1}{9}$, $s_{\ell_i} = \frac{1}{9}$, $s_n = 0$ for $n_i \leq n < m_{i+1}$, which proves the lemma.

Definition. Given a matrix $A = (a_{n\nu})$, a function $0 < F(n) \uparrow \infty$ $(n=0,1,\ldots)$ is called a summability function for A when $s_n \to 0$ (A) holds whenever $s_n = O(1)$ and $A(n,s) \leq F(n)$.

If $A = C_1$, then every $F(n) = o(n)$ is a summability function. In fact, let $|s_n| \leq K$, then

$$|s_0 + \ldots + s_n| \leq KA(n,s) \leq KF(n) = o(n)$$

which implies $s_n \to 0$ (C_1) .

Theorem II. 33. ([66]) Let A be regular, and let F(n) be a summability function. Then, $a_n = O(\frac{1}{F(n)})$ is not a Tauberian condition for A .

Proof. We have $\Sigma \frac{1}{F(n)} = \infty$ since otherwise $\frac{n}{F(n)} = o(1)$, which implies $F(n) \geq n$ for large n . It would follow that $\{s_n\}$ with $s_n = 1$ for large n exists such that $A(n,s) \leq F(n)$, and this would imply $s_n \to 0$ (A) - but A is regular. By Lemma II. 10. there is a divergent and A summable series Σa_n with $a_n = O(\frac{1}{F(n)})$. This proves the theorem.

As an example let us consider the method C_1 . If $0 < \lambda_n \to \infty$ and $F(n) = \sup_{k \leq n} \frac{k+1}{\lambda_k}$, then $0 < F(n) \uparrow \infty$, $F(n) = o(n)$, and it follows from $\frac{1}{F(n)} \leq \frac{\lambda_{n+1}}{n+1}$ that $a_n = O(\frac{\lambda_{n+1}}{n+1})$ is not a Tauberian condition for C_1 (for a result in the opposite direction see page 6).

The next theorem is useful to determine summability functions for a given method.

Theorem II, 34. ([66]) Let $A = (a_{n\nu})$ be regular. If $0 < f(n) \uparrow \infty$, and if $\sum_{\nu=0}^{\infty} |a_{n\nu} - a_{n,\nu+1}| f(\nu) = O(1)$ $(n \to \infty)$, then every $0 < F(n) \uparrow \infty$, $F(n) = o(f(n))$ is a summability function for A .

Proof. Let $s_n = O(1)$, $A(n,s) = o(f(n))$, then $S_n = s_0 + \ldots + s_n = o(f(n))$.

Moreover, as $k \to \infty$,

$$o(1) = \sum_{\nu=k}^{\infty} |a_{n\nu}-a_{n,\nu+1}| f(\nu) \geq f(k) \sum_{\nu=k}^{\infty} |a_{n\nu}-a_{n,\nu+1}| \geq f(k)|a_{nk}| .$$

But

$$\sum_{\nu=0}^{k} a_{n\nu} s_\nu = \sum_{\nu=0}^{k-1} (a_{n\nu}-a_{n,\nu+1}) S_\nu + a_{nk} S_k ,$$

and this implies

$$\sum_{\nu=0}^{\infty} a_{n\nu} s_\nu = \sum_{\nu=0}^{\infty} (a_{n\nu}-a_{n,\nu+1}) S_\nu ,$$

from which it follows that

$$\left|\sum_{\nu=0}^{\infty} a_{n\nu} s_\nu\right| \leq \sum_{\nu=0}^{N} |a_{n\nu}-a_{n,\nu+1}| |S_\nu| + \varepsilon \sum_{\nu=0}^{\infty} |a_{n\nu}-a_{n,\nu+1}| f(\nu)$$

i.e. $s_n \to 0$ (A) .

Examples.

1. Consider M_p with $0 < p_n \uparrow$, $\frac{P_n}{p_n} \uparrow \infty$. For $f(n) = \frac{P_n}{p_n}$ it follows that

$$\sum_{\nu=0}^{\infty} |a_{n\nu}-a_{n,\nu+1}| f(\nu) = \frac{1}{P_n} \sum_{\nu=0}^{n-1} |p_\nu - p_{\nu+1}| \frac{P_\nu}{p_\nu} +1 \leq \frac{1}{P_n} \frac{P_n}{p_n} (p_n - p_0) +1 \leq 2 .$$

Theorems II. 33. and II. 34. show that the Tauberian condition $a_n = O(\frac{p_n}{P_n})$ which we obtained earlier (page 46) is best possible.

2. For Euler's method E_1 every $0 < F(n) \uparrow \infty$, $F(n) = o(\sqrt{n})$ is a summability function. This can be shown in the following way. We first observe that

$$2^{-n} \sum_{\nu=0}^{n} \binom{n}{\nu}(n-2\nu)^2 = n . \tag{43}$$

Next

$$\frac{1}{2^n} \sum_{\nu=0}^{n} \left|\binom{n}{\nu} - \binom{n}{\nu+1}\right| \sqrt{\nu+1} = \frac{1}{2^n} \sum_{\nu=0}^{n} \binom{n}{\nu} \frac{|n-2\nu-1|}{\sqrt{\nu+1}} .$$

But

$$\sum_{\nu=0}^{n} \binom{n}{\nu} \frac{|n-2\nu|}{\sqrt{\nu+1}} = \left\{ \sum_{|n-2\nu| \leq \sqrt{n}} + \sum_{\sqrt{n} < |n-2\nu| \leq \frac{3}{4}n} + \sum_{\frac{3}{4}n < |n-2\nu| \leq n} \right\} \binom{n}{\nu} \frac{|n-2\nu|}{\sqrt{\nu+1}}$$

$$= O(1) \sum_{\nu=0}^{n} \binom{n}{\nu} \frac{\sqrt{n}}{\sqrt{n+1}} + O(1) \sum_{\nu=0}^{n} \binom{n}{\nu} \frac{|n-2\nu|^2}{n} + O(1) \sum_{\nu=0}^{n} \binom{n}{\nu} \frac{|n-2\nu|^2}{n} = O(2^n) .$$

Remarks.

1. There are cases where Theorem II. 34. does not give all summability functions. A simple proof shows, for instance, that the only summability functions which follow from Theorem II. 34. for N_p, $p_n = \frac{1}{n+1}$, are $F(n) = o(\log n)$. On the other hand, every function $0 < F(n) \uparrow \infty$, $F(n) = O(n^{\varepsilon_n})$, $0 < \varepsilon_n \to 0$ is a summability function for this method. This follows for $|s_n| \leq K$, $A(n, s) \leq F(n)$ from

$$\frac{1}{\log n} \left| \sum_{\nu=0}^{n} \frac{s_\nu}{n+1-\nu} \right| \leq \frac{K}{\log n} \sum_{\nu=n-F(n)}^{n} \frac{1}{n+1-\nu} = O(1) \frac{\log F(n)}{\log n} = o(1) .$$

(Observe that $\frac{1}{n+1-\nu} \uparrow$ for $\nu \uparrow$.)

2. We have seen (page 46) that $a_n = O(\frac{1}{n \log n})$ is a Tauberian condition for M_p, $p_n = \frac{1}{n+1}$. Theorem II. 33. does not show that this condition is best possible. Otherwise $F(n) = n \sqrt{\log n}$, for instance, would have to be a summability function which is clearly impossible (every bounded sequence would be limitable to zero). More generally, the methods developed so far cannot handle Tauberian conditions of the type $a_n = O(\frac{1}{F(n)})$ with $F(n) \geq n$. In what follows we discuss another approach to the question of best Tauberian conditions.

Lemma II. 11. ([65]) If $A = (a_{n\nu})$ satisfies the conditions

$$\sum_{\nu=0}^{\infty} |a_{n\nu}| < \infty , \qquad \eta_n = \sup_\nu |a_{n\nu}| \to 0 \qquad (n \to \infty) ,$$

then there exists a divergent sequence $\{s_n\}$ consisting of elements 0 and 1 only such that $s_n \to 0$ (A).

Proof. Select a sequence $0 < n_1 < n_2 < \ldots$ such that $\eta_n \leq \frac{1}{2^i}$ for $n \geq n_i$. Next, select a sequence $0 < \nu_2 < \nu_3 < \ldots$ such that

$$\sum_{\nu=\nu_{i+1}}^{\infty} |a_{n\nu}| \leq \frac{1}{2^i} \qquad \text{for} \qquad n_i \leq n < n_{i+1} .$$

If $s_{\nu_i} = 1$, $s_\nu = 0$ $\nu \neq \nu_i$, then for $n_k \leq n < n_{k+1}$, $k \geq 2$,

$$|\sigma_n| = \left| \sum_{i=2}^{\infty} a_{n\nu_i} \right| \leq \sum_{i=2}^{k} |a_{n\nu_i}| + \sum_{i=k+1}^{\infty} |a_{n\nu_i}| \leq k\eta_n + \sum_{\nu=\nu_{k+1}}^{\infty} |a_{n\nu}| \leq \frac{k}{2^k} + \frac{1}{2^k} \to 0 .$$

Theorem II. 35. ([67]) Let $A = (a_{n\nu})$ be regular and assume that for a sequence $0 < \nu_1 < \nu_2 < \ldots$

$$\sup_k \sum_{\nu=\nu_k}^{\nu_{k+1}-1} |a_{n\nu}| \to 0 \qquad (n \to \infty) .$$

Then, if $c_\nu \geq 0$, $c_\nu \to 0$, $\sum_{\nu=\nu_k}^{\nu_{k+1}-1} c_\nu \geq \delta > 0$, it follows that $a_n = O(c_n)$ is not a Tauberian condition for A.

Proof. Let $b_{nk} = \sum_{\nu_k \leq \nu < \nu_{k+1}} |a_{n\nu}|$, then $B = (b_{nk})$ satisfies the assumptions of Lemma II. 11.. Thus, there is a sequence $\{s_k\}$, $s_{k(i)} = 1$, $s_k = 0$, $\nu \nmid k(i)$, such that

$$T_n = \sum_{i=1}^{\infty} b_{n,k(i)} = \sum_{i=1}^{\infty} \sum_{\nu=\nu_{k(i)}}^{\nu_{k(i)+1}-1} |a_{n\nu}| \to 0 \qquad (n \to \infty) .$$

There is a sequence $\{\mu_i\}$, $\nu_{k(i)} < \mu_i < \nu_{k(i)+1}$, and there are numbers $0 < \alpha_i < 1$, $-1 < \beta_i < 0$ such that

$$\sum_{\nu_{k(i)} \leq \nu \leq \mu_i} \alpha_i c_\nu = \frac{\delta}{3} \quad , \quad \sum_{\mu_i < \nu \leq \nu_{k(i)+1}-1} \beta_i c_\nu = -\frac{\delta}{3}$$

(c.f. the proof of Lemma II. 10.).

If $a_\nu = \alpha_i c_\nu$, $\nu_{k(i)} \leq \nu \leq \mu_i$, $a_\nu = \beta_i c_\nu$, $\mu_i < \nu < \nu_{k(i)+1}$, $a_\nu = 0$ otherwise, then $|s_n| = |a_0 + \ldots + a_n| \leq \frac{\delta}{3}$, $\{s_n\}$ is divergent, $a_n = O(c_n)$ and

$$\left| \sum_{\nu=0}^{\infty} a_{n\nu} s_\nu \right| \leq \frac{\delta}{3} T_n \to 0 .$$

Problems.

1. Show that $a_n = O(\frac{\lambda_n}{\sqrt{n}})$, $\lambda_n \to \infty$, is not a Tauberian condition for E_1.

2. Show that every function $0 < F(n) \uparrow \infty$, $F(n) = o(\sqrt{n})$, is a summability function for E_p, $p > 0$.

3. Show that every function $0 < F(n) \uparrow \infty$, $F(n) = o(n)$, is a summability function for Abel's method A. (Abel's method is not a matrix method of the type considered in Theorems II. 33. and II. 34., but it is obvious that these theorems are still true for this method; see also the remarks on page 24.)

4. Show that every function $0 < F(n) \uparrow \infty$, $F(n) = o(\sqrt{n})$, is a summability function for Borel's method.

5. Let $\lambda_\nu = o(\nu \log \nu)$, $\lambda_\nu \to \infty$, $c_\nu = \frac{\lambda_\nu}{\nu \log \nu}$, and choose integers ν_k with the properties $\lambda_{\nu_1} > 1$, $\nu_{k+1} > \nu_k$,

$$\frac{\lambda_{\nu_k}}{\lambda_{\nu_k}-1} \log \nu_k \leq \log \nu_{k+1} \leq \frac{\lambda_{\nu_k}}{\lambda_{\nu_k}-1} \log \nu_k + 1 \quad .$$

Show as an application of Theorem II. 35. that the Tauberian condition $a_n = O(\frac{1}{n \log n})$ for M_p, $p_n = \frac{1}{n+1}$, is best possible.

Chapter III Tauberian Theorems

In this chapter we will discuss Tauberian Theorems which cannot be obtained using a mean value theorem.

1. The M. Riesz convexity theorem

Given $\{s_\nu\}$, we write $S_n^\alpha(s_\nu) = S_n^\alpha = \sum_{\nu=0}^{n} A_{n-\nu}^{\alpha-1} s_\nu$ (cf. chapter I, 2.).

Theorem III. 1. (see M. Riesz [91]) Let $0 \leq \alpha < \gamma < \beta$ and let

$$S_n^\alpha = O(U(n)) \quad , \quad S_n^\beta = O(V(n)) \quad , \quad 0 < U(n), V(n) \uparrow \infty$$

then

$$S_n^\gamma = O(1)\, U^{\frac{\beta-\gamma}{\beta-\alpha}}(n)\, V^{\frac{\gamma-\alpha}{\beta-\alpha}}(n)$$

If $S_n^{\alpha} = o(U)$ or $S_n^{\beta} = o(V)$, then o also holds in the conclusion. For the proof we require two lemmas.

Lemma III. 1. Let $0 < \gamma < \beta \leq 1$, then a number $C(\gamma, \beta)$ exists such that

$$\sup_{\nu \leq n} |S_\nu^\gamma| \leq C(\gamma,\beta) \left(\sup_{\nu\leq n} |s_\nu|\right)^{\frac{\beta-\gamma}{\beta}} \left(\sup_{\nu \leq n} |S_\nu^\beta|^{\frac{\gamma}{\beta}}\right) .$$

Proof. $$S_n^\gamma = \sum_{\nu=0}^{k} A_{n-\nu}^{\gamma-1} s_\nu + \sum_{\nu=k+1}^{n} A_{n-\nu}^{\gamma-1} s_\nu$$

($-1 \leq k \leq n$; for $k = -1$ or $k = n$ the first or second sum is zero). Partial summation yields

$$S_n^\gamma = \sum_{\nu=0}^{k-1} \Delta_\nu \left(\frac{A_{n-\nu}^{\gamma-1}}{A_{n-\nu}^{\beta-1}}\right) \sum_{\mu=0}^{\nu} A_{n-\mu}^{\beta-1} s_\mu + \frac{A_{n-k}^{\gamma-1}}{A_{n-k}^{\beta-1}} \sum_{\mu=0}^{k} A_{n-\mu}^{\beta-1} s_\mu + \sum_{\nu=k+1}^{n} A_{n-\nu}^{\gamma-1} s_\nu .$$

It follows (cf. (13) page 35) that

$$(1) \qquad |S_n^\gamma| \leq 2 \frac{A_{n-k}^{\gamma-1}}{A_{n-k}^{\beta-1}} \sup_{\nu\leq n} |S_\nu^\beta| + A_{n-k-1}^{\gamma} \sup_{\nu \leq n} |s_\nu|$$

(observe that $\dfrac{A_{n-\nu}^{\gamma-1}}{A_{n-\nu}^{\beta-1}} \uparrow$ as $\nu \uparrow$).

Let $\sup_{\nu\leq n} |s_\nu| = A_n$, $\sup_{\nu \leq n} |S_\nu^\beta| = B_n$, and we may assume that $A_n > 0$ since $A_n = 0$ implies $S_n^\gamma = 0$.

a) If $1 \leq \left(\frac{B_n}{A_n}\right)^{\frac{1}{\beta}} \leq n$, then take k with

$$0 \leq n - \left(\frac{B_n}{A_n}\right)^{\frac{1}{\beta}} \leq k \leq (n+1) - \left(\frac{B_n}{A_n}\right)^{\frac{1}{\beta}} \leq n .$$

It follows from (1) that

$$|S_n^\gamma| \leq D(\beta,\gamma) \left\{(n+1-k)^{\gamma-\beta} B_n + (n-k)^\gamma A_n\right\} ,$$

and this implies

$$|S_n^\gamma| \leq D(\beta,\gamma) \left\{B_n \left(\frac{B_n}{A_n}\right)^{\frac{\gamma-\beta}{\beta}} + A_n \left(\frac{B_n}{A_n}\right)^{\frac{\gamma}{\beta}}\right\} = 2\, D(\beta,\gamma)\, A_n^{\frac{\beta-\gamma}{\beta}} B_n^{\frac{\gamma}{\beta}} .$$

b) If $\frac{B_n}{A_n} < 1$ then take $k = n$. It follows from (1) that

$$|S_n^\gamma| \leq 2 B_n \leq 2 A_n^{\frac{\beta-\gamma}{\beta}} B_n^{\frac{\gamma}{\beta}} .$$

c) If $(\frac{B_n}{A_n})^{\frac{1}{\beta}} > n$, then $k = -1$. It follows from (1) that

$$|S_n^\gamma| \leq D(\beta,\gamma)(\frac{B_n}{A_n})^{\frac{\gamma}{\beta}} A_n = D(\beta,\gamma) A_n^{\frac{\beta-\gamma}{\beta}} B_n^{\frac{\gamma}{\beta}} .$$

This proves the lemma.

Lemma III. 2. Let $0 \leq \alpha < \gamma < \beta$. Then $C(\alpha,\beta,\gamma)$ exists such that

$$\sup_{\nu \leq n} |S_\nu^\gamma| \leq C(\alpha,\beta,\gamma)(\sup_{\nu\leq n} |S_\nu^\alpha|)^{\frac{\beta-\gamma}{\beta-\alpha}} (\sup_{\nu\leq n} |S_\nu^\beta|)^{\frac{\gamma-\alpha}{\beta-\alpha}} .$$

Proof. We proceed by induction and assume that the lemma is true for $\beta - a \leq 2^k$ $(k=0,1,2,\ldots)$. Lemma III. 1. states that $k = 0$ is permissible here (replace in Lemma III. 1. s_ν by S_ν^α, γ by $\gamma - \alpha$, β by $\beta - \alpha$ and observe that $S^\beta(S^\alpha) = S^{\alpha+\beta}$). Let $2^k < \beta - \alpha \leq 2^{k+1}$, and write $\gamma_i = \alpha + \frac{i}{4}(\beta-\alpha)$, $i=0,1,2,3,4$, $\sup_{\nu\leq n} |S_\nu^{\gamma_i}| = A_n(i)$. It follows from $\gamma_{i+2} - \gamma_i \leq 2^k$ (Lemma III. 2. is assumed to be true for $\beta - \alpha \leq 2^k$!) that

$$A_n(1) \leq C_1 \sqrt{A_n(0) A_n(2)}$$
$$A_n(2) \leq C_2 \sqrt{A_n(1) A_n(3)}$$
$$A_n(3) \leq C_3 \sqrt{A_n(2) A_n(4)}$$

and this implies

$$A_n(2) \leq C_1 C_2^2 C_3 \sqrt{A_n(0) A_n(4)}$$

It follows that for $\alpha < \gamma < \gamma_2$ (note that $\gamma_2 - \alpha \leq 2^k$)

$$\sup_{\nu\leq n} |S_\nu^\gamma| \leq C(\alpha,\gamma_2,\gamma) A_n(0)^{\frac{\gamma_2-\gamma}{\gamma_2-\alpha}} A_n(2)^{\frac{\gamma-\alpha}{\gamma_2-\alpha}}$$

$$\leq C(\alpha,\gamma_2,\gamma) A_n(0)^{\frac{\gamma_2-\gamma}{\gamma_2-\alpha}} (C_1C_2^2C_3)^{\frac{\gamma-\alpha}{\gamma_2-\alpha}} A_n(0)^{\frac{\gamma-\alpha}{2(\gamma_2-\alpha)}} A_n(4)^{\frac{\gamma-\alpha}{2(\gamma_2-\alpha)}}$$

$$= C(\alpha,\beta,\gamma)\, A_n(0)^{\frac{\beta-\gamma}{\beta-\alpha}} A_n(4)^{\frac{\gamma-\alpha}{\beta-\alpha}} ,$$

and similarly for $\gamma_2 < \gamma < \beta$. This proves Lemma III. 2.

Theorem III. 1. is an immediate consequence of Lemma III. 2.

Applications of Theorem III. 1. Let $S_n^{\beta}(s_\nu) = o(n^{\beta})$ ($\beta > 0$), i.e., $s_n \to 0$ (C_β) and let $s_n = O(1)$. Then $S_n^{\gamma}(s_\nu) = o(n^{\gamma})$ for every $0 < \gamma < \beta$, i.e., $s_n \to 0$ (C_γ). We thus have proven

Theorem III. 2. (see [27], [4], [90]) All Cesàro methods of positive order are equivalent for bounded sequences.

If $a_n = O(\frac{1}{n})$ and $\{s_n\} \in C_\alpha$, then $s_n = O(1)$. For a proof we may assume that $\alpha \geq 1$, and then we have

$$\sigma_n - s_n = \frac{1}{A_n^{\alpha}} \sum_{\nu=0}^{n} (A_{n-\nu}^{\alpha} - A_n^{\alpha}) a_\nu = \frac{O(1)}{A_n^{\alpha}} \sum_{\nu=1}^{n} \frac{1}{\nu+1} \sum_{\mu=n-\nu+1}^{n} A_\mu^{\alpha-1} = O(n^{-\alpha}) \sum_{\nu=1}^{n} A_n^{\alpha-1} = O(1)$$

(compare this with problem 3 page 7).

This result, combined with Theorems I. 2. and III. 2. , shows that $a_n = O(\frac{1}{n})$ is a Tauberian condition for every C_α , a result which shows that Theorem III. 1. is also true for $\alpha = -1$, $U(n) = \frac{1}{n+1}$, $\beta > \gamma > 0$, $V(n) = n^{\beta}$ (note that $S_n^{-1}(s_\nu) = s_n - s_{n-1} = a_n$). For further results see for instance [11], [15].

Problems.

1. Let $f(x)$ be defined for $x > 0$ and twice differentiable. Show that $f(x) = O(U(x))$, $f'' = O(V(x))$ ($0 < U(x), V(x) \uparrow \infty$, for $x \to \infty$) implies $f'(x) = O(\sqrt{U(x)V(x)})$.
2. Show that $a_n = O(\frac{\lambda_n}{n})$, $\lambda_n \to \infty$, is not a Tauberian condition for C_α .
3. Find a sequence $\{s_n\}$ with the following properties : $s_n = o(n)$, $s_n \to 0$ (C_2), $s_n \not\to 0$ (C_1).

2. Tauberian theorems for the methods of Abel, Euler and Borel - first discussion

We first prove Tauber's original theorem ([101]).

Theorem III. 3. If $s_n \to s$ (Abel) and $a_n = o(\frac{1}{n})$, then $s_n \to s$

Proof. Let $x = 1 - \frac{1}{n}$ and $\sigma(x) = \sum_0^\infty x^\nu a_\nu$. We have

$$\sigma(x) - s_n = -\sum_{\nu=0}^{n} (1-x^\nu)a_\nu + \sum_{\nu=n+1}^{\infty} a_\nu x^\nu = I + II .$$

But

$$1 - x^\nu = (1-x)(1+x+\dots+x^{\nu-1}) \le \nu(1-x) = \frac{\nu}{n}$$

and

$$|I| \le \frac{1}{n} \sum_{\nu=0}^{n} |\nu a_\nu| = o(1) \quad \text{(by Lemma 0. 1.)}.$$

Furthermore,
$$II = o(\tfrac{1}{n}) \sum_{\nu=n+1}^{\infty} x^\nu = o(\tfrac{1}{n}) \frac{x^{n+1}}{1-x} = o(1) .$$

This completes the proof.

Remarks.

1. We have already mentioned in chapter I that theorems of a structure similar to that of Theorem III. 3. are called Tauberian theorems. Abel's limit theorem states that the Abel method is regular, and regularity statements for individual methods are frequently called Abelian theorems.

2. If $a_n = O(\frac{1}{n})$, then the proof of Theorem III. 3. shows that $s_n \to s$ (A) implies $s_n = O(1)$. If we combine this with Theorem III. 2. and Frobenius' theorem (page 25, then we have a new proof of the fact that $s_n \to s$ (C_α) and $a_n = O(\frac{1}{n})$ imply $s_n = O(1)$ (see the end of the last section).

3. In the next section we will prove a theorem of J. E. Littlewood which states that $a_n = O(\frac{1}{n})$ is a Tauberian condition for Abel's method. We have seen in chapter II. 10. that $a_n = O(\frac{\lambda_n}{n})$, $\lambda_n \to \infty$, is not a Tauberian condition for C_1, and hence not for Abel's method since $C_1 \subsetneqq A$.

Next we turn to Euler's method.

Theorem III. 4. (see [53]) If $s_n \to s$ (E_1) and $a_n = o(\frac{1}{\sqrt{n}})$, then $s_n \to s$.

Proof. We have $s_n = o(\sqrt{n})$ and $|s_n - s_\nu| = o(1) \frac{|n-\nu|}{\sqrt{n}}$ for $n \to \infty$ and uniformly in ν (consider the cases $\nu \le \frac{n}{2}$ and $\nu > \frac{n}{2}$). Next

$$\sigma_{2n} - s_n = \frac{1}{2^{2n}} \sum_{\nu=0}^{2n} \binom{2n}{\nu}(s_\nu - s_n) = o\left(\frac{1}{\sqrt{n}} \frac{1}{2^{2n}}\right) \sum_{\nu=0}^{2n} \binom{2n}{\nu} |n-\nu| = o(1)$$

(we have shown on page 60 that $\frac{1}{2^{2n}} \sum_{\nu=0}^{2n} \binom{2n}{\nu} \frac{|n-\nu|}{\sqrt{\nu+1}} = O(1)$), and this proves the theorem.

Remark. If $s_n \to s$ (E_1) and $a_n = O(\frac{1}{\sqrt{n}})$, then $s_n = O(1)$.

Concerning Borel's method we prove

Theorem III. 5. ([26]) If $s_n \to s$ (B) and $a_n = o(\frac{1}{\sqrt{n}})$, then $s_n \to s$.

Proof. We show first that $e^{-x} \sum_{\nu=0}^{\infty} \frac{x^\nu}{\nu!} \frac{|x-\nu|}{\sqrt{\nu+1}} = O(1)$ $(x \to \infty)$. In order to prove this we observe that

$$e^{-x} \sum_{\nu=0}^{\infty} \frac{x^\nu}{\nu!} (x-\nu)^2 = x$$

and it follows that

$$\sum_{\nu=0}^{\infty} \frac{x^\nu}{\nu!} |\nu-x| \frac{1}{\sqrt{\nu+1}} = \left\{ \sum_{|\nu-x| \le \sqrt{x}} + \sum_{\sqrt{x} < |\nu-x| \le \frac{3}{4}x} + \sum_{\frac{3}{4}x < |\nu-x|} \right\} \frac{x^\nu}{\nu!} |\nu-x| \frac{1}{\sqrt{\nu+1}}$$

$$= O(1) \sum_{\nu=0}^{\infty} \frac{x^\nu}{\nu!} + O(1) \sum_{\nu=0}^{\infty} \frac{x^\nu}{\nu!} \frac{|\nu-x|^2}{x} + O(1) \sum_{\nu=0}^{\infty} \frac{x^\nu}{\nu!} \frac{|\nu-x|^2}{x} = O(e^x) .$$

As in the proof of Theorem III. 4., we have $|s_n - s_\nu| = o(1) \frac{|n-\nu|}{\sqrt{n}}$. Next, for $x = n$,

$$\sigma(x) - s_n = o(\frac{1}{\sqrt{n}})\, e^{-x} \sum_{\nu=0}^{\infty} \frac{x^\nu}{\nu!} |n-\nu| ,$$

and $\sigma(x) - s_n \to 0$ follows from

$$e^{-x} \sum_{\nu \ge 2x}^{\infty} \frac{x^\nu}{\nu!} |x-\nu| \le e^{-x} \sum_{\nu=0}^{\infty} \frac{x^\nu}{\nu!} \frac{|x-\nu|^2}{x} = O(1) .$$

Remarks.

1. If $s_n \to s$ (B) and $a_n = O(\frac{1}{\sqrt{n}})$, then $s_n = O(1)$.

2. Because of $E_1 \subsetneqq B$ (see page 25) Theorem III. 4. is a consequence of Theorem III. 5. .

3. In section 7 of this chapter we will prove that $a_n = O(\frac{1}{\sqrt{n}})$ is a Tauberian condition for B . We have seen in chapter II. 10. that $a_n = O(\frac{\lambda_n}{n})$, $\lambda_n \to \infty$, is not a Tauberian condition for E_1 , and hence not for B since $E_1 \subsetneqq B$.

We have proven in this section three so-called o-Tauberian theorems. Each of these theorems shows that from the corresponding O-Tauberian condition we obtain at least $s_n = O(1)$. There is a general theorem by T. Vijayaraghavan ([104], [105]; see also [25]) which leads from O-Tauberian conditions to $s_n = O(1)$ for a wide class of methods (including those discussed here).

Problems.

1. Show that Theorem III. 3. remains true when $n a_n \to 0$ is replaced by $n a_n \to 0 \ (C_1)$.
2. Given a sequence $0 < \lambda_0 < \lambda_1 < \ldots \to \infty$, a series Σa_n is called summable (A, λ) to s if $f(x) = \sum_0^\infty a_n e^{-\lambda_n x}$ exists for $x > 0$, and if $f(x) \to s$ for $x \to +0$. Show that the method (A, λ) is regular.
3. Prove the following Tauberian theorem for (A, λ)-summability: If $\Sigma a_n = s \ (A, \lambda)$ and $a_n = o(\frac{\lambda_{n+1} - \lambda_n}{\lambda_{n+1}})$, then $\Sigma a_n = s$.

3. Littlewood's theorem for Abel summability

Theorem III. 6. If $a_n \geq 0$ and $(1-x) \sum_{\nu=0}^{\infty} a_\nu x^\nu \to c$ as $x \to 1-0$ (the series should converge for $|x| < 1$), then $\frac{s_n}{n} \to c \quad (n \to \infty)$.

For the proof we require the following

Lemma III. 3. Let $g(t) = 0 \quad (0 \leq t < \frac{1}{e})$, $g(t) = \frac{1}{t} \quad (\frac{1}{e} \leq t \leq 1)$, then given $\varepsilon > 0$, there are polynomials $p(t)$, $P(t)$ with

$$p(t) \leq g(t) \leq P(t) \quad , \quad (0 \leq t \leq 1) \, , \qquad \int_0^1 (P(t) - p(t))\, dt \leq \varepsilon$$

We omit the simple proof (use the Weierstrass approximation theorem).

Proof of Theorem III. 6. It follows from

$$(1-x)\sum_{n=0}^{\infty} a_n x^n x^{kn} = \frac{1-x}{1-x^{k+1}}(1-x^{k+1})\sum_{n=0}^{\infty} a_n (x^{k+1})^n \to \frac{c}{k+1}$$

that $(1-x)\sum_{n=0}^{\infty} a_n x^n Q(x^n) \to c\int_0^1 Q(t)\,dt$ for every polynomial Q. Using this result and Lemma III. 3. we find

$$c\int_0^1 p(t)dt \leq \liminf_{x\to 1}(1-x)\sum_{n=0}^{\infty} a_n x^n g(x^n) \leq \limsup_{x\to 1}(1-x)\sum_{n=0}^{\infty} a_n x^n g(x^n) \leq c\int_0^1 P(t)dt .$$

We have $\int_0^1 p(t)\,dt = \int_0^1 g(t)\,dt - \int_0^1 (g(t)-p(t))\,dt \geq 1-\varepsilon$ and similarly $\int_0^1 P(t)\,dt \leq 1+\varepsilon$. This implies $(1-x)\sum_{n=0}^{\infty} a_n x^n g(x^n) \to c$, and the proof of the theorem is completed when we observe that for $x = 1-\frac{1}{m}$

$$(1-x)\sum_{0}^{\infty} a_n x^n g(x^n) = \frac{1}{m}\sum_{(1-\frac{1}{m})^n \geq \frac{1}{e}} a_n = \frac{1}{m}\sum_{n=0}^{m-1} a_n$$

(note that $(1-\frac{1}{m})^m < \frac{1}{e} < (1-\frac{1}{m})^{m-1}$).

Theorem III. 7. If $a_n \geq -K$, $(1-x)\sum_{\nu=0}^{\infty} a_\nu x^\nu \to c$ $(x\to 1-0)$, then $\frac{s_n}{n} \to c$.

Proof. $(1-x)\sum_{\nu=0}^{\infty}(a_\nu + K)x^\nu \to (c+K)$ implies by Theorem III. 6. that $\frac{s_n + (n+1)K}{n} \to c+K$.

Theorem III. 8. If $s_n \geq -K$, $s_n \to s$ (Abel), then $s_n \to s$ (C_1).

Proof. Theorem III. 7. with s_n instead of a_n.

Theorem III. 8. implies that Abel- and C_1-summability are equivalent for bounded sequences, and it follows from Theorem III. 2. that Abel- and C_α -summability $(\alpha > 0)$ are equivalent for bounded sequences.

Now let $s_n \to s$ (Abel) and $a_n = O(\frac{1}{n})$; it follows that $s_n = O(1)$ (see remark 2 after Theorem III. 3.), hence we have $s_n \to s$ (C_1) which implies $s_n \to s$ because of Theorem I. 2.. We thus have proven

Theorem III. 9. (J. E. Littlewood) If $s_n \to s$ (Abel) and $a_n = O(\frac{1}{n})$, then $s_n \to s$.

(J. E. Littlewood proved this theorem in 1910, [64]. The proof presented here goes back to J. Karamata [49]; for another proof see W. Jurkat [41]. Theorem III. 9. can also be obtained from N. Wiener's theory which will be discussed in sections 6 and 7 of this chapter.)

Problems.

1. Show that $\sum_{\nu=0}^{n} |\nu a_\nu|^p = O(n)$ is for every $p > 1$ a Tauberian condition for Abel's method (see [99]).

2. Show that slow oscillation is a Tauberian condition for Abel's method.

3. Show that Theorem III. 6. is not true without the assumption $a_n \geq 0$.

4. Analytic continuation - gap Tauberian theorems

Let $f(z)$ regular in a region D, and for a neighborhood of $z_0 \in D$ let $f(z) = \sum_0^\infty a_n(z-z_0)$. The region of convergence of this series is - disregarding boundary points - of the type $|z-z_0| < r$ for some $0 < r \leq \infty$. Wherever the series is convergent it is also Borel-summable, and the question arises whether it might be B-summable in a sizable larger region (see problem 1. on page 25).

We begin a discussion of this problem in greater generality for a regular matrix method $A = (a_{n\nu})$. We have (assume that $z_0 = 0$)

$$s_n(z) = \sum_{\nu=0}^{n} a_\nu z^\nu = \frac{1}{2\pi i} \int_C \frac{f(\zeta)}{\zeta - z} \left(1 - \left(\frac{z}{\zeta}\right)^{n+1}\right) d\zeta$$

($f(z)$ regular inside of C, z and $z_0 = 0$ inside of C).

But

$$\sigma_n(z) = \sum_{\nu=0}^{\infty} a_{n\nu} s_\nu(z) = \frac{1}{2\pi i} \int_C \frac{f(\zeta)}{\zeta - z} d\zeta \sum_{\nu=0}^{\infty} a_{n\nu} - \frac{z}{2\pi i} \int_C \frac{f(\zeta)}{\zeta(\zeta - z)} \sum_{\nu=0}^{\infty} a_{n\nu} \left(\frac{z}{\zeta}\right)^\nu d\zeta = I + II,$$

provided that the interchange of $\int$ and Σ is justifiable.

Because of $I = f(z) \sum_{\nu=0}^{\infty} a_{n\nu} \to f(z)$ it turns out that $\sigma_n(z) \to f(z)$ if $II \to 0$, and sufficient for this is that a C can be found such that $\sum_{\nu=0}^{\infty} a_{n\nu} (\frac{z}{\zeta})^{\nu} \to 0$ uniformly for every $\zeta \in C$.

We now discuss the special case of Borel - summability - it does not matter that we have here a matrix of the type $a_{x,\nu} = e^{-x} \frac{x^{\nu}}{\nu!}$. Here (the inversion of $\int$ and Σ is easily justified)

$$\sum_{\nu=0}^{\infty} a_{x\nu} (\tfrac{z}{\zeta})^{\nu} = e^{x(\frac{z}{\zeta}-1)} \to 0 \quad \text{for} \quad x \to \infty \quad \text{iff} \quad \operatorname{Re} \frac{z}{\zeta} < 1 \quad .$$

A simple discussion shows that $\operatorname{Re} \frac{z}{\zeta} < 1$ is true exactly for the points ζ outside of the closed disc B_z with center $\frac{z}{2}$ and radius $|\frac{z}{2}|$. Thus, B-summability takes place for all points z with the property that $B_z \subset D$ (we are then able to construct C - as required - outside B_z).

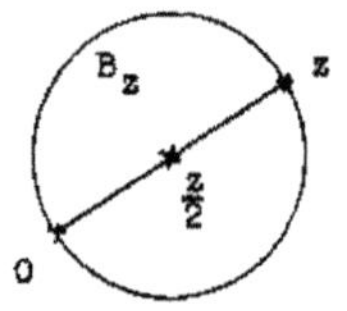

Theorem III. 10. (see [8]) Let $f(z)$ be regular in a domain D with $0 \in D$, $f(z) = \sum_{0}^{\infty} a_n z^n$ (in a neighborhood of 0). Then, this series is B summable to $f(z)$ on the point set

$$B = \bigcup_{B_z \subset D} z$$

There is another way to describe B. Given $\zeta \neq 0$, let H_{ζ} be the open half plane (of complex numbers z) $\operatorname{Re} \frac{z}{\zeta} < 1$.

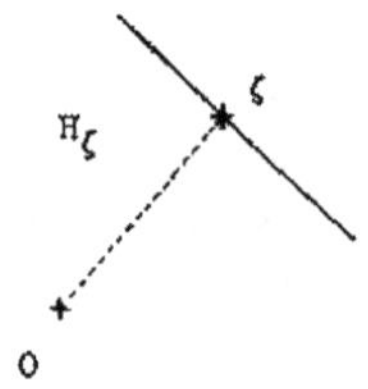

Then we have $B = \bigcap_{\zeta \notin D} H_{\zeta}$ (and B is a starshaped set).

In fact, $z \in H_{\zeta}$ iff $\operatorname{Re} \frac{z}{\zeta} < 1$ and $\operatorname{Re} \frac{z}{\zeta} < 1$ iff $\zeta \notin B_z$. Thus, if D^* is the complement of D, then $z \in H_{\zeta}$ for every $\zeta \in D^*$ iff $D^* \cap B_z = \emptyset$, i.e., $B_z \subset D$.

Example. If $f(z) = \frac{1}{1-z^4}$, then $B = H_1 \cap H_i \cap H_{-1} \cap H_{-i}$, i.e., B is the square $-1 < \operatorname{Re} z$, $\operatorname{I} z < +1$.

For E_1 summability we have

$$\sum_{\nu=0}^{n} a_{n\nu} (\tfrac{z}{\zeta})^{\nu} = \left(\frac{1+\frac{z}{\zeta}}{2}\right)^{n} \to 0 \quad \text{for} \quad n \to \infty \quad \text{iff} \quad |1 + \tfrac{z}{\zeta}| < 2 \quad .$$

A simple discussion shows that $|1+\frac{z}{\zeta}| < 2$ is true exactly for the points outside of the closed disc E_z with center $\frac{z}{3}$ and radius $|\frac{2}{3}z|$.

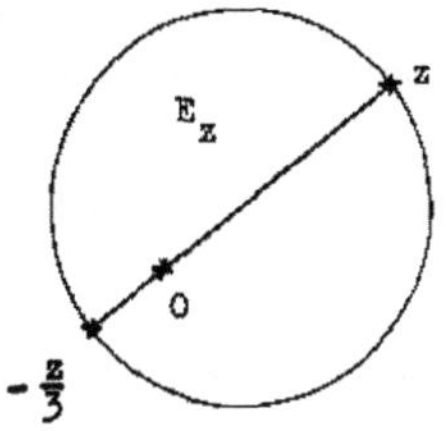

Thus we have

<u>Theorem III. 11.</u> Assumptions as in Theorem III. 10. . Then $\Sigma\, a_n a^n$ is E_1 - summable to $f(z)$ for all points $E = \bigcup_{E_z \subset D} z$.

There is a generalization of Theorems III. 10. and III. 11. due to Y. Okada [76] and G. G. Lorentz [68] for matrix methods in general. We will mention it without proof.

Given $f(z)$, regular at $z = 0$, we define the <u>Mittag-Leffler star</u> S_f of f as the set

$$S_f = \bigcup_{0 \le \phi \le 2\pi} \left\{ re^{i\phi} \; ; \; f(z) \text{ regular for } z = \rho e^{i\phi} \; , \; 0 \le \rho \le r \right\} .$$

Let $A = (a_{n\nu})$ be a regular matrix method, and let R , $\{|z| < 1\} \subset R$, be an open starshaped region such that $\Sigma\, z^n$, $z \in R$, is summable A to $\frac{1}{1-z}$, and uniformly summable on every compact subset of R . Let $C(R)$ be the complement of R , and write R' for the image of $C(R)$ under the mapping $1/z$.

<u>Theorem III. 12.</u> If S_f is the Mittag-Leffler star of f , then its power series expansion $\Sigma\, a_n z^n$ at zero is summable A to $f(z)$ for

$$z \in S_A^f = \bigcup_{\zeta R' \subset S_f} \zeta \qquad (\zeta R' = \zeta w : w \in R') .$$

Another representation of S_A^f is

$$S_A^f = \bigcap_{\zeta \notin S_f} \zeta R \qquad (\zeta R = \zeta w : w \in R) .$$

(For a proof see [68] .)

We now combine results on analytic continuation with a theorem of Tauberian nature and obtain results in complex function theory.

A series $\Sigma\, a_n$ is called a gap series if $0 < n_1 < n_1' < n_2 < n_2' < \dots$ exists such that

$$\frac{n_k'}{n_k} \geq 1+\lambda > 1 \qquad \text{and} \qquad a_n = 0 \qquad \text{for} \qquad n_k < n \leq n_k' .$$

To illustrate a "gap"- Tauberian theorem, we discuss first the case of C_1 .

Theorem III. 13. Let $\Sigma\, a_n$ be a gap series and suppose that $\Sigma\, a_n = 0 \quad (C_1)$; then $s_{n_k} \to 0$.

Instead of a Tauberian order condition we have here a "gap-condition" .

Proof.

$$\sigma_{n_k'} = \frac{1}{n_k'+1} \sum_{0 \leq \nu < n_k} s_\nu + \frac{1}{n_k'+1} \sum_{\nu=n_k}^{n_k'} s_\nu = \frac{n_k}{n_k'+1}\, \sigma_{n_k-1} + s_{n_k} \frac{n_k'-n_k+1}{n_k'+1} .$$

Therefore,

$$o(1) = o(1) + s_{n_k} \frac{1-\frac{n_k-1}{n_k'}}{1+\frac{1}{n_k'}} ,$$

and this implies $s_{n_k} \to 0$.

We prove next a "gap" - Tauberian theorem for E_1 - summability.

Lemma III. 4. For $0 < \varepsilon < 1$:

$$\phi(\varepsilon) = (1-\varepsilon)^{1-\varepsilon}(1+\varepsilon)^{1+\varepsilon} > 1 .$$

Proof. $\phi(0) = 1$, $\phi'(\varepsilon) = \phi(\varepsilon)\log\frac{1+\varepsilon}{1-\varepsilon} > 0$.

Theorem III. 14. (cf. [115]) Given $0 < n_1 < n_1' < n_2 < n_2' < \ldots$, $\frac{n_k'}{n_k} \geq 1+\lambda > 1$, there exists a $\delta = \delta(\lambda) > 0$ such that $s_{n_k} \to s$ for every gap series $\Sigma\, a_n = s(E_1)$ ($a_n = 0$ for $n_k < n \leq n_k'$) with $s_n = a_0 + \ldots + a_n = O((1+\delta)^n)$.

Proof. Choose m_k and $0 < \varepsilon < 1$ such that

$$n_k \leq m_k(1-\varepsilon) \leq m_k(1+\varepsilon) \leq n_k' .$$

Then

$$\sigma_{2m_k} = \frac{1}{2^{2m_k}} \sum_{m_k(1-\varepsilon) \leq \nu \leq m_k(1+\varepsilon)} \binom{2m_k}{\nu} s_{n_k} + \frac{1}{2^{2m_k}} \sum_{\text{Rest}} \binom{2m_k}{\nu} s_\nu = \text{I} + \text{II} .$$

Here,
$$II = O\left((1+\delta)^{2m_k}\right) \frac{1}{2^{2m_k}} \sum_{\text{Rest}} \binom{2m_k}{\nu}$$

It follows from Stirling's formula by a straightforward calculation that

$$\binom{2m_k}{[m_k(1-\varepsilon)]} = O(1) \frac{2^{2m_k}}{\left\{(1-\varepsilon)^{1-\varepsilon}(1+\varepsilon)^{1+\varepsilon}\right\}^{m_k}\sqrt{m_k}} ,$$

and this implies that $s_{n_k} \to s$ if $(1+\delta)^2 < (1-\varepsilon)^{1-\varepsilon}(1+\varepsilon)^{1+\varepsilon}$.

Remarks. Let Σa_n be a series with $a_n = 0$ for $n \neq n_k$, $n_{k+1} - n_k \geq \lambda n_k$, $\lambda > 0$. It follows from Theorem III. 13. that $\Sigma a_n = s$ (C_1) implies $\Sigma a_n = s$. There is an extension of this result by Hardy and Littlewood [30] : It remains true when C_1 is replaced by Abel's method (for further results see e. g. [12]).

Concerning Borel's method D. Gaier [19] proved in 1965 the following gap Tauberian theorem (for earlier results see the bibliography in [19]): If $\Sigma a_n = s$ (B) , $a_n = 0$ for $n \neq n_k$, $n_{k+1} - n_k > \lambda\sqrt{n_k}$, $\lambda > 0$, then $\Sigma a_n = s$.

As an application of Theorems III. 11. and III. 14. we obtain Ostrowski's theorem on over-convergence (see [77], [78]).

Theorem III. 15. Let $f(z) = \sum_0^\infty a_n z^n$ have radius of convergence 1 , and assume that $f(z)$ is regular for $z = 1$. Then, if Σa_n is a gap series it follows that $s_{n_k}(z) = \sum_0^{n_k} a_n z^n \to f(z)$ in some (circular) neighborhood of $z = 1$.

Proof. It follows from Theorem III. 11. that $\Sigma a_n z^n$ is E_1 - summable to $f(z)$ in some neighborhood U_1 of $z = 1$. Furthermore, $a_n = O((1+\eta)^n)$ for every $\eta > 0$, and $s_n(z) = O(1)\left\{(1+\eta)|z|\right\}^n$ for every $\eta > 0$ and $|z| \geq 1$. Therefore, by Theorem III. 14. there is a neighborhood $U_2 \subset U_1$ of $z = 1$ such that $s_{n_k}(z) \to f(z)$ in U_2 .

A consequence of Theorem III. 15. is J. Hadamard's gap theorem .

Theorem III. 16. Assume that $0 < n_1 < n_2 < \dots$, $\frac{n_{k+1}}{n_k} \geq 1 + \lambda > 1$.

Then $|z| = 1$ is a natural boundary for $f(z) = \Sigma a_n z^n$ if $a_n = 0$ for $n \neq n_k$, and if the radius of convergence is 1 .

Proof. $\Sigma a_n z^n$ is a gap series, if we write $n_k' = n_{k+1} - 1$. If $f(z)$ is regular for $z = 1$, then $\sum_0^{n_k} a_n z^n$ converges in a neighborhood U of $z = 1$, and this means, that $\sum_0^\infty a_n z^n$

converges in U, which is impossible. An obvious modification shows that also $z = e^{i\phi}$ for every $\phi \neq 0$ is not a regular point.

Problems.

1. Determine the point set where Σz^{np} $(p=1,2,\ldots)$ is summable B.
2. Let B be the so-called Borel-star defined in Theorem III. 10.. Show that $\Sigma a_n z^n$ is not summable B for $z \in C(\bar{B})$, the complement of the closure of B.
3. Show that Theorem III. 13. remains true when C_1 is replaced by a normal, regular and positive matrix A with $M_K(A)$ and $\liminf_{n\to\infty} \sum_{(1-\varepsilon)n \leq \nu \leq n} a_{n\nu} > 0$ for every $\varepsilon > 0$.

5. Some properties of Borel summability

Theorem III. 17. If $\Sigma s_n x^n$ exists for $0 \leq x < 1$, and if $s_n \to s$ (B), then $s_n \to s$ (Abel).

Proof. Using the notation $B(x) = e^{-x} \sum_{\nu=0}^{\infty} \frac{x^\nu}{\nu!} s_\nu$, we have

$$(1-x) \sum_{\nu=0}^{\infty} s_\nu x^\nu = (1-x) \int_0^\infty e^{t(x-1)} B(xt)\,dt \qquad (0 < x < 1)$$

(replacing s_n by $s_n + (1+\varepsilon)^n$ the interchange of Σ and $\int$ can be justified by using Fubini's theorem on series with positive terms). But a Toeplitz-type argument (sufficient part) shows that $B(x) \to s$ $(x \to \infty)$ implies $(1-x) \sum_0^\infty s_\nu x^\nu \to s$.

(There is a related result by Hardy and Littlewood [26]: If $\Sigma a_n = s$ (B) and $a_n = o(n^\rho)$, $\rho \geq -1/2$, then $\Sigma a_n = s$ $(C_{2\rho+1})$; see also Theorem III. 5..)

The fact that $B \subsetneqq A$ if A is applicable is not surprising if we recall the Tauberian conditions for these methods. A diagram similar to that on page 7 looks as follows :

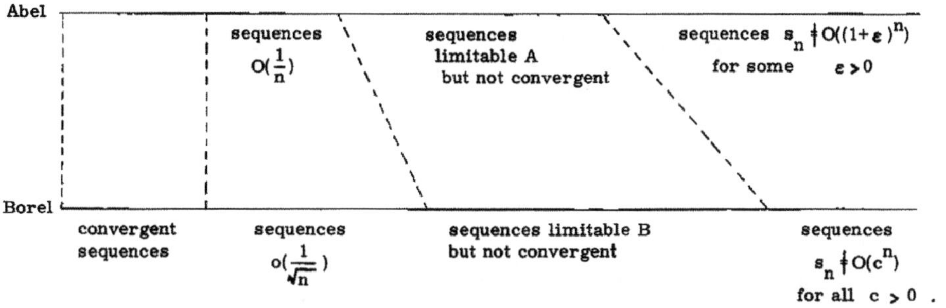

Thus, it appears that B is useful for rapidly divergent sequences (as in the case of analytic extension) but A is better for sequences which are close to convergence. The situation is similar for C_1 and E_1 .

In section 7 of the present chapter we will prove that $a_n = O(\frac{1}{\sqrt{n}})$ is a Tauberian condition for B , and in view of remark 1. page 68 it is sufficient to consider only bounded sequences. In what follows we are going to discuss a modification of Borel's method for those sequences.

<u>Lemma III. 5.</u> Let $\frac{1}{2} < \zeta < \frac{2}{3}$, $|\nu - x| \leq x^{\zeta}$, then (as $x \to \infty$ and uniformly in ν)

$$e^{-x}\frac{x^{\nu}}{\nu!} = \sqrt{\frac{1}{2\pi x}}\, e^{-\frac{1}{2x}(\nu-x)^2}\left(1 + O\left(\frac{|\nu-x|+1}{x}\right) + O\left(\frac{|\nu-x|^3}{x^2}\right)\right) .$$

<u>Proof</u> by Stirling's formula

$$\log e^{-x}\frac{x^{\nu}}{\nu!} = -x + \nu + (x - \nu - x - \tfrac{1}{2}) \log \frac{\nu}{x} - \tfrac{1}{2}\log 2\pi x + O(\tfrac{1}{\nu}) ,$$

and the result follows if we write $\log\frac{\nu}{x} = \frac{\nu-x}{x} - \frac{1}{2}\left(\frac{\nu-x}{x}\right)^2 + O\left(\left(\frac{\nu-x}{x}\right)^3\right)$.

We have in particular

$$(2) \qquad e^{-x}\frac{x^{\nu}}{\nu!} = \sqrt{\frac{1}{2\pi x}}\left\{\exp - \frac{1}{2x}(\nu - x)^2\right\} + o\left(e^{-x}\frac{x^{\nu}}{\nu!}\right) .$$

<u>Lemma III. 6.</u> Assumptions as in Lemma III. 5. . Then we have uniformly in ν for $x \to \infty$,

(3) $$e^{-\frac{1}{2x}(\nu-x)^2} = \int_{\nu}^{\nu+1} e^{-\frac{1}{2x}(t-x)^2} dt + o(\sqrt{x}\ \ e^{-x}\frac{x^{\nu}}{\nu!})$$

(4) $$x^{-\frac{1}{2}} e^{-\frac{1}{2x}(\nu-x)^2} = \int_{\nu}^{\nu+1} e^{-2(\sqrt{t}-\sqrt{x})^2} \frac{dt}{\sqrt{t}} + o(e^{-x}\frac{x^{\nu}}{\nu!})$$

Proof.

$$\int_{\nu}^{\nu+1} e^{-\frac{1}{2x}(t-x)^2} dt = e^{-\frac{1}{2x}(\nu-x)^2} \int_{\nu}^{\nu+1} e^{-\frac{t-\nu}{2x}(t-\nu+2(\nu-x))} dt = e^{-\frac{1}{2x}(\nu-x)^2}(1+o(1))$$

(observe that $|\nu - x| \leq x^{\zeta}$). This proves (3).

In order to prove (4) we use the identity

$$(a-b)^2 = \frac{(a^2-b^2)^2}{4a^2} + \frac{(a^2-b^2)^3}{4a^4}\left((\frac{a}{a+b})^2 + 2(\frac{a}{a+b})^3 \right) .$$

It follows that

$$\int_{\nu}^{\nu+1} e^{-2(\sqrt{t}-\sqrt{x})^2} \frac{dt}{\sqrt{t}} = \int_{\nu}^{\nu+1} \frac{1}{\sqrt{x}} e^{-\frac{1}{2x}(t-x)^2} \left\{ \exp \frac{(t-x)^3}{2x^2}\left((\frac{\sqrt{x}}{\sqrt{x}+\sqrt{t}})^2 + 2(\frac{\sqrt{x}}{\sqrt{x}+\sqrt{t}})^3 \right) \right\} \frac{\sqrt{x}}{\sqrt{t}} dt$$

$$= \int_{\nu}^{\nu+1} e^{-\frac{1}{2x}(t-x)^2} (1+o(1)) \frac{dt}{\sqrt{x}} = \frac{(1+o(1))}{\sqrt{x}} \int_{\nu}^{\nu+1} e^{-\frac{1}{2x}(t-x)^2} dt .$$

This proves (4) .

Theorem III. 18. If $s_n \to s$ (B) $s_n = O(1)$, and if $s(t) = s_{\nu}$ $(\nu \leq t < \nu+1)$, then , for $x \to \infty$,

(5) $$\sqrt{\frac{1}{2\pi x}} \int_0^{\infty} e^{-\frac{1}{2x}(t-x)^2} s(t) dt \to s$$

and

(6) $$\sqrt{\frac{1}{2\pi}} \int_0^{\infty} e^{-2(\sqrt{t}-\sqrt{x})^2} s(t)\frac{dt}{\sqrt{t}} \to s .$$

Proof. If $\zeta > \frac{1}{2}$, then

$$e^{-x} \sum_{\nu\geq 0, |\nu-x|\geq x^{\zeta}} \frac{x^{\nu}}{\nu!} \leq \frac{e^{-x}}{x^{2\zeta}} \sum_{\nu=0}^{\infty} \frac{x^{\nu}}{\nu!}(\nu-x)^2 = \frac{1}{x^{2\zeta-1}} \to 0$$

(see the proof of Theorem III. 5.). It follows that

$$e^{-x} \sum_{|\nu-x|\leq x^{\zeta}} \frac{x^{\nu}}{\nu!} s_{\nu} \to s \qquad ,$$

and we find from (2), (3), and (4) that for $\frac{1}{2} < \zeta < \frac{2}{3}$ and $x \to \infty$

$$\sqrt{\frac{1}{2\pi x}} \int_{|t-x|\leq x^{\zeta}} e^{-\frac{1}{2x}(t-x)^2} s(t)dt \to s$$

$$\sqrt{\frac{1}{2\pi}} \int_{|t-x|\leq x^{\zeta}} e^{-2(\sqrt{t}-\sqrt{x})^2} s(t)\frac{dt}{\sqrt{t}} \to s \qquad .$$

The proof is completed when we observe that

$$x^{-\frac{1}{2}} \int_{|t-x|\geq x^{\zeta}} e^{-\frac{1}{2x}(t-x)^2} dt = \int_{|v|\geq x^{\zeta-\frac{1}{2}}} e^{-\frac{1}{2}v^2} dv \to 0$$

and similarly

$$\int_{\substack{|t-x|\geq x^{\zeta} \\ t\geq 0}} e^{-2(\sqrt{t}-\sqrt{x})^2} \frac{dt}{\sqrt{t}} \to 0 \qquad .$$

Problems.

1. Show that neither $A \subsetneqq B$ nor $B \subsetneqq A$ is true. Similarly for C_{α} and E_p
2. Show that Theorem III. 18. is true when $s_n = O(1)$ is replaced by $s_n = o(\sqrt{n})$.

6. Wiener's Tauberian theorems

For proofs of the basic theorems of this theory see e. g. G. H. Hardy [25]. (These theorems were originally proved by N. Wiener [107] in 1932.)

Theorem III. 19. (Hardy [25] Theorem 232) If $g(t), h(t) \in L(0,\infty)$, and if

$$(7) \qquad \int_0^\infty g(t)\, t^{ix}\, dt \neq 0 \qquad (-\infty < x < \infty),$$

then $s(t) = O(1)$ (s(t) real and measurable) and

$$\frac{1}{x}\int_0^\infty g(\frac{t}{x}) s(t)\, dt \to 0 \quad (x \to \infty) \qquad \text{imply} \qquad \frac{1}{x}\int_0^\infty h(\frac{t}{x}) s(t)\, dt \to 0 \quad (x \to \infty).$$

A real function $s(t)$, measurable in $(0,\infty)$, is called slowly oscillating if for every $\varepsilon > 0$ a $\delta = \delta(\varepsilon)$ and $T = T(\varepsilon)$ exists such that $|s(y) - s(x)| \leq \varepsilon$ whenever $x \geq T$ and $1 \leq \frac{y}{x} \leq 1 + \delta$. The function $s(t)$ is called slowly decreasing, if for every $\varepsilon > 0$ a $\delta = \delta(\varepsilon)$ and $T = T(\varepsilon)$ exists such that $s(y) - s(x) \geq -\varepsilon$ whenever $x \geq T$ and $1 \leq \frac{y}{x} \leq 1 + \delta$.

Theorem III. 20. (Hardy [25] Theorem 233) If $g(t) \in L(0,\infty)$, and if (7) holds, then $s(t) = O(1)$, $\frac{1}{x}\int_0^\infty g(\frac{t}{x}) s(t)\, dt \to 0$ $(x \to \infty)$ and $s(t)$ slowly decreasing imply $s(t) \to 0$.

7. Applications of Wiener's theory

We discuss things first in some detail for the C_1 method.

Let $s_n \to s$ (C_1). Then it follows with $s(t) = s_\nu$ $\nu \leq t < \nu + 1$, that for $n \leq x < n+1$ $(n=1,2,3,\ldots)$

$$\frac{1}{x}\int_0^x s(t)dt = \frac{n}{x}\,\frac{s_0+\ldots+s_{n-1}}{n} + \frac{x-n}{x}\, s_n \to s$$

(since $s_n = o(n)$ by Theorem I. 1.).

Next, we note that $s_n \to s$ (C_1) and $a_n = O(\frac{1}{n})$ imply $s_n = O(1)$ (see problem 3. page 7).
If $g(t) = 1$ $(0 \leq t \leq 1)$, $g(t) = 0$ $(t > 1)$, then $g(t) \in L(0, \infty)$ and

$$\int_0^\infty g(t)\, t^{ix}\, dt = \frac{1}{1+ix} \neq 0 \qquad \text{for} \qquad -\infty < x < \infty .$$

It follows from $\frac{1}{x}\int_0^\infty g(\frac{t}{x}) s(t)\, dt = \frac{1}{x}\int_0^x s(t)\, dt$ by Theorem III. 20. that $s_n \to s$ (C_1) and $a_n = O(\frac{1}{n})$ imply $s_n \to s$.

Remarks. As an application of Theorem III. 20. we have just proven Theorem I. 2. with the Tauberian condition $a_n = O(\frac{1}{n})$. It is not difficult to extend this proof to the Tauberian condition $\liminf (s_m - s_n) \geq 0$ for $1 \leq \frac{m}{n} \to 1$ $(n \to \infty)$. In a similar way we can prove Littlewood's theorem. Moreover, it follows from Theorem III. 19. that all Cesáro methods, and Abel's method, are equivalent for bounded sequences.

Next, we turn to Borel's method. If $s_n \to 0$ (B) and $a_n = O(\frac{1}{\sqrt{n}})$, then $s_n = O(1)$ and

$$\int_0^\infty e^{-2(u-\sqrt{y})^2}\, s(u^2) du \to 0 \qquad y \to \infty , \quad s(u) = \sum_{\nu \leq u} a_\nu .$$

(See remark 1. page 68 , and Theorem III. 18. .)

Writing $\sqrt{y} = \log x$, $u = \log t$ we have for $x \to \infty$

$$\frac{1}{x}\int_0^\infty e^{-2(\log \frac{t}{x})^2}\, \frac{x}{t}\, f(t) dt \to 0 ,$$

where $f(t) = s((\log t)^2)$ $(f(0) = 0$ for $0 \leq t < 1)$.

Here, $f(t) = O(1)$ and for $y > x > 1$

$$f(y) - f(x) = \sum_{(\log x)^2 < \nu \leq (\log y)^2} a_\nu = O(\frac{1}{\log x}) \left\{ (\log y)^2 - (\log x)^2 + 1 \right\}$$

$$= O(\frac{1}{\log x}) + O(\log \frac{y}{x} \frac{\log xy}{\log x}) \to 0 \qquad \frac{y}{x} \to 1 \qquad (x \to \infty) .$$

Next $\frac{1}{t} e^{-2(\log t)^2} \in L(0, \infty)$ and

$$\int_0^\infty t^{ix} g(t) dt = \int_{-\infty}^{+\infty} e^{-2v^2 + vix} dv = e^{-\frac{x^2}{8}} \int_{-\infty}^{+\infty} e^{-2(v - \frac{ix}{4})^2} dv = \sqrt{\frac{\pi}{2}}\, e^{-\frac{x^2}{8}} \neq 0 \quad (-\infty < x < \infty) .$$

From Theorem III. 20. we obtain

Theorem III. 21. (Hardy-Littlewood [28]) $s_n \to s$ (B) and $a_n = O(\frac{1}{\sqrt{n}})$ imply $s_n \to s$.

Remark. The condition $a_n = O(\frac{1}{\sqrt{n}})$ can be replaced by more general ones ($\{s_n\}$ slowly oscillating or slowly decreasing), see for instance [14].

Problem.

Use Theorem III. 20. to show that $a_n = O(\frac{1}{n})$ is a Tauberian condition for every method C_α.

8. Lambert's method

A series $\sum_1^\infty a_n$ is called summable to s by the Lambert method L if

$$\sigma(x) = (1-x) \sum_{\nu=1}^{\infty} \frac{\nu a_\nu x^\nu}{1-x^\nu} \to s \qquad x \to 1-0 ,$$

assuming that the series converges for $|x| < 1$ which is true iff $a_n = O((1+\varepsilon)^n)$ for every $\varepsilon > 0$ (see [3], [51], [108]).

If we write $x = e^{-\frac{1}{y}}$ $(y > 0)$, $s(t) = \sum_{\nu \le t} a_\nu$ $(a_0 = 0)$, $g(t) = \frac{te^{-t}}{1-e^{-t}}$, then $\sum a_\nu$ is summable L to s iff (note that $(1-x) \approx \frac{1}{y}$)

$$\frac{1}{y}\int_0^\infty \frac{t\, e^{-\frac{t}{y}}}{1-e^{-\frac{t}{y}}}\, ds(t) = -\int_0^\infty s(t)\, dg(\tfrac{t}{y}) = -\frac{1}{y}\int_0^\infty g'(\tfrac{t}{y}) s(t)\, dt \to s \qquad y \to \infty .$$

Furthermore, $g'(t) = \frac{e^{-t}(1-t-e^{-t})}{(1-e^{-t})^2} \leq 0$ for $t \geq 0$, and

$$\int_0^\infty |dg(\tfrac{t}{y})| = -\int_0^\infty dg(\tfrac{t}{y}) = g(0) - g(\infty) = 1 \;, \quad \int_0^T |dg(\tfrac{t}{y})| = g(0) - g(\tfrac{T}{y}) \to 0 \qquad y \to \infty .$$

This implies that L is regular. Thus, we have

Theorem III. 22. The method L is regular.

Next, we prove

Theorem III. 23. If $\Sigma a_n = s(L)$ and $a_n = o(\frac{1}{n})$, then $\Sigma a_n = s$.

Proof. Let $x = 1 - \frac{1}{n}$ $(n = 1, 2, \dots)$. Then

$$\sigma(x) - s_n = (1-x) \sum_{\nu=n+1}^{\infty} \frac{\nu a_\nu x^\nu}{1-x^\nu} + (1-x) \sum_{\nu=1}^{n} a_\nu \left(\frac{\nu x^\nu}{1-x^\nu} - n\right) = I + II$$

But

$$I = o(1-x) \sum_{\nu=n+1}^{\infty} \frac{x^\nu}{1-x^\nu} = o(1-x) \frac{x^{n+1}}{1-x^{n+1}} \frac{1}{1-x} = o(1) .$$

Next

$$n - \frac{\nu x^\nu}{1-x^\nu} = \nu + \frac{1}{1-x^\nu} \left(\frac{1-x^\nu}{1-x} - \nu\right) \leq \nu$$

and

$$n - \frac{\nu x^\nu}{1-x^\nu} = \frac{1}{1-x^\nu} \left(\frac{1-x^\nu}{1-x} - \nu x^\nu\right) \geq 0 .$$

It follows that

$$|II| \leq \frac{1}{n} \sum_{\nu=1}^{n} |\nu a_\nu| \to 0 .$$

Remark. This proof shows that $s_n \to s$ (L) and $a_n = O(\frac{1}{n})$ imply $s_n = O(1)$.

We wish to show next that $a_n = O(\frac{1}{n})$ is a Tauberian condition. In order to apply Wiener's theory we must show that (7) of Theorem III. 19. holds. But for $\varepsilon > 0$

$$-\int_0^\infty t^{ix+\varepsilon} g'(t)dt = (ix+\varepsilon) \int_0^\infty t^{ix+\varepsilon-1} g(t)dt = (ix+\varepsilon) \sum_{\nu=0}^{\infty} \int_0^\infty t^{ix+\varepsilon} e^{-(\nu+1)t} dt$$

$$= (ix+\varepsilon)\ \Gamma(1+\varepsilon+ix) \sum_{\nu=0}^{\infty} \frac{1}{(\nu+1)^{1+\varepsilon+ix}}$$

i.e.,

$$-\int_0^{\infty} t^{ix} g'(t)dt = \Gamma(1+ix) \lim_{\varepsilon\to 0} (ix+\varepsilon)\ \zeta(1+\varepsilon+ix)$$

where ζ is Riemann's zeta function. This function has a simple pole at 1 and is $\neq 0$ on the line $\operatorname{Re} z = 1$. Thus, we have

Theorem III. 24. $s_n \to s$ (L) and $a_n = O(\frac{1}{n})$ imply $s_n \to s$.

Remark. A stronger theorem is true; namely, $L \subsetneqq$ Abel (see [29]), which implies Theorem III. 24.. The proof of this theorem requires more complicated facts on the ζ function.

Appendix. For the sake of completeness we will give a proof that $\zeta(1+ix) \neq 0$ for x real (J. Hadamard and C. de la Vallée Poussin 1896).

If $s = \sigma + it$, and if $\sigma > 1$, then $\zeta(s) = \sum_1^{\infty} \frac{1}{n^s}$. A short calculation shows that for $\sigma > 1$

$$\sum_1^{\infty} \frac{(-1)^{n-1}}{n^s} = (1-2^{1-s})\ \zeta(s) .$$

The series on the left is convergent for $\sigma > 0$. Hence, we have an analytic continuation of $\zeta(s)$ onto $\operatorname{Re} s > 0$ with the exception of a pole of the first order at $s = 1$. Next, for $\sigma > 1$,

$$(8) \qquad -\zeta'(s) = \zeta(s) \sum_1^{\infty} \frac{\Lambda_n}{n^s} ,$$

where $\Lambda_n = \log p$ if $n = p^{\alpha}$ ($\alpha = 1, 2, \dots$, p prime), $\Lambda_n = 0$ otherwise. In fact, (multiplication of Dirichlet series) if $n = p_1^{\alpha_1} \dots p_k^{\alpha_k}$ then

$$\sum_{d|n} \Lambda_d = \alpha_1 \log p_1 + \dots + \alpha_k \log p_k = \log n$$

(d runs through the divisors of n, and we have to consider only $d = p_1, p_1^2, \dots, p_1^{\alpha_1}, \dots, p_k^{\alpha_k}$). Formula (8) implies $\zeta(\sigma+it) \neq 0$ for $\sigma > 1$. If $\zeta(s) = (s-\hat{s})^{\alpha}\ g(s)$, $\operatorname{Re} \hat{s} = 1$, $g(\hat{s}) \neq 0$ and regular in a neighborhood of $\hat{s}$, then

$$(s-\hat{s}) \frac{\zeta'(s)}{\zeta(s)} \to \alpha \qquad s \to \hat{s} .$$

Let $t \neq 0$ and let $s_0 = 1$, $s_1 = 1+it$, $s_2 = 1+2it$; using (8) we have for $\sigma > 1$

$$H = \mathrm{Re}\left((\sigma-s_0)3\frac{\zeta'(\sigma)}{\zeta(\sigma)} + (\sigma+it-s_1)4\frac{\zeta'(\sigma+it)}{\zeta(\sigma+it)} + (\sigma+2it-s_2)\frac{\zeta'(\sigma+2it)}{\zeta(\sigma+2it)}\right)$$

$$= -(\sigma-1)\sum_1^\infty \frac{\Lambda_n}{n^\sigma}(3+4\cos(t\log n)+\cos(2t\log n)) \leq 0$$

because of $3+4\cos x + \cos 2x = 2(1+\cos x)^2 \geq 0$. But $H \to -3+4\beta+\gamma$ for $\sigma \to 1$ (observe that $\alpha = -1$ at $\hat{s} = 1$), if ζ has a zero of order β , γ at s_1, s_2 , and $H \leq 0$ implies $\beta = 0$.

9. The prime number theorem

1. The Möbius formula.

Let
$$\mu(n) = \begin{cases} 1 & \text{for } n=1 \\ (-1)^k & \text{for } n = p_1p_2\cdots p_k , \quad p_i \text{ prime}, \quad p_i \neq p_j , \\ 0 & \text{otherwise} \end{cases}$$

then

(9) $$\sum_{d|n}\mu(d) = 0 \qquad \text{for} \qquad n > 1 .$$

(We omit the simple proof.)

Given $G(x)$ for $x > 0$, let $F(x) = \sum_{n\leq x} G(\frac{x}{n})$ ($n \geq 1$ an integer), then $G(x) = \sum_{n\leq x}\mu(n)F(\frac{x}{n})$ (the proof follows immediately from (9)). Likewise, if $F(n) = \sum_{d|n} G(\frac{n}{d})$, then $G(n) = \sum_{d|n} \mu(d) F(\frac{n}{d})$.

2. We have $\sum_{1}^{\infty} \frac{\mu(n)}{n} = 0$. This follows from the O-Tauberian theorem for Lambert summability if $\sum \frac{\mu(n)}{n} = 0 \quad (L)$. But

$$(1-x) \sum_{n=1}^{\infty} \frac{\mu(n) x^n}{1-x^n} = (1-x) \sum_{n=1}^{\infty} \mu(n) \sum_{k=0}^{\infty} x^{n(k+1)} = (1-x) \sum_{m=1}^{\infty} x^m \sum_{n|m} \mu(n) = x(1-x) \quad .$$

A consequence is (by partial summation)

$$(10) \qquad \sum_{\nu \le n} \mu(\nu) = o(n) \quad ,$$

which follows with the notation $m(t) = \sum_{1 \le \nu \le t} \frac{\mu(\nu)}{\nu}$ from

$$\sum_{\nu \le n} \mu(\nu) = \int_{1-0}^{n} t\,dm(t) = n\,m(n) - \int_{1}^{n} m(t)\,dt \quad .$$

3. With the Euler-Mascheroni constant C which occurs for example in the asymptotic formula

$$\sum_{k \le x} \frac{1}{k} = \log x + C + O(\tfrac{1}{x})$$

we set

$$\chi(x) = \sum_{k \le x} \left\{ \psi(\tfrac{x}{k}) - \tfrac{x}{k} + \log \tfrac{x}{k} + C \right\} \quad , \quad \psi(x) = \sum_{n \le x} \Lambda_n$$

Möbius formula yields

$$(11) \qquad \psi(x) - x + \log x + C = \sum_{d \le x} \chi(\tfrac{x}{d}) \mu(d) \quad .$$

From $\log n = \sum_{d|n} \Lambda_d$ (page 84) it follows that

$$\sum_{n \le x} \log n = \sum_{n \le x} \sum_{kd=n} \Lambda_d = \sum_{k \le x} \sum_{d \le \frac{x}{k}} \Lambda_d = \sum_{k \le x} \psi(\tfrac{x}{k}) \quad .$$

Therefore, we obtain

$$\chi(x) = \sum_{n \le x} \log n - x \left\{ \log x + C + O(\tfrac{1}{x}) \right\} + [x] \log x - \sum_{k \le x} \log k + [x]\, C \quad ,$$

i. e.

$$(12) \qquad \chi(x) = O(\log (x+1))$$

4. Next, we need <u>Axer's theorem</u> (this may be considered as a theorem of Abelian nature, see [25]):

If $\sum_{1 \leq \nu \leq x} a_\nu = o(x)$, $a_n = O(1)$, $\chi \in V(1,T)$ for every T , $\chi(x) = O(x^\alpha)$ for some $0 < \alpha < 1$, then $\sum_{1 \leq \nu \leq x} \chi(\frac{x}{\nu}) a_\nu = o(x)$.

<u>Proof.</u> Let $0 < \delta < 1$, then

$$\sum_{1 \leq \nu \leq \delta x} \chi(\tfrac{x}{\nu}) a_\nu = O(x^\alpha)\, \delta^{1-\alpha} x^{1-\alpha} = O(x\delta^{1-\alpha}) .$$

Assume that $m-1 < \delta x \leq m$, $N \leq x < N+1$, (m and N integers) ; then

$$\sum_{\delta x \leq \nu \leq x} \chi(\tfrac{x}{\nu}) a_\nu = \sum_{\nu=m}^{N-1} \{\chi(\tfrac{x}{\nu}) - \chi(\tfrac{x}{\nu+1})\}\, s_\nu + \chi(\tfrac{x}{N})\, s_N - \chi(\tfrac{x}{m})\, s_{m-1}$$

$$= o(x) \int_{\delta x}^{x} |d\chi(\tfrac{x}{t})| + o(x) = o(x) \int_{1}^{1/\delta} |d\chi(t)| + o(x) .$$

This completes the proof (make δ small).

It now follows from (10) , (11), (12) and Axer's theorem that $\psi(x) - x = o(x)$.

5. Let $\vartheta(x) = \sum_{p \leq x} \log p$ (p prime) , then $\vartheta(x) \leq \psi(x) = O(x)$. Furthermore,

$$\psi(x) = \vartheta(x) + \vartheta(\sqrt{x}) + \dots + \vartheta(\sqrt[k]{x})$$

for every $k > \frac{\log x}{\log 2}$

It follows that

$$\psi(x) = \vartheta(x) + O(1) \frac{\log x}{\log 2} \sqrt{x} ,$$

and this implies

(13) $$\vartheta(x) = x + o(x) .$$

6. Let $\pi(x) = \sum_{p<x} 1$ (p prime ; i.e., $\pi(x)$ is the number of primes $< x$).

We have

$$\pi(x) = \int_{3/2}^{x} \frac{1}{\log t}\, d\vartheta(t) = \frac{\vartheta(x)}{\log x} + \int_{3/2}^{x} \frac{\vartheta(t)}{t(\log t)^2}\, dt = \frac{\vartheta(x)}{\log x} + O\left(\frac{x}{(\log x)^2}\right)$$

(note that $\vartheta(x) = O(x)$).

Using (13) we obtain the prime number theorem (J. Hadamard, Ch. de la Vallée Poussin, 1896):

Theorem III. 25.

$$\frac{\pi(x) \log x}{x} \to 1 \qquad \text{as} \qquad x \to \infty .$$

Chapter IV. Hausdorff and Nörlund summability

1. The consistency of Hausdorff and Nörlund methods

In chapter II. 3. we have seen that all regular Hausdorff methods are consistent, and this is also true for all regular and real Nörlund methods. One might ask whether also Hausdorff and Nörlund methods are consistent (see [103]). It is the purpose of this section to show that this is the case.

We call a sequence $\{s_n\}$ limitable to s by the method A^* (extended Abel method) if $\Sigma s_n z^n$ has a positive radius of convergence and defines an analytic function which is regular for $0 < x < 1$, and if

$$\sigma(x) = (1-x)\sum_{n=0}^{\infty} s_n x^n \to s \qquad x \to 1-0 \quad .$$

Lemma IV. 1. $A^* \subsetneqq A^* H$ (and identical limits) for every regular Hausdorff method H. (This is an extension of a theorem of O. Szasz [100]).

Proof. Let $s_n \to s \ (A^*)$, and assume that $H = H_\chi = \Delta (d_n) \Delta$, $d_n = \int_0^1 t^n d\chi(t)$. We write $f(z) = \sigma(\frac{1}{1+z}) = z \sum_{\nu=0}^{\infty} \frac{s_\nu}{(1+z)^{\nu+1}}$, and $f(z)$ is regular for $|z| \geq R$ and $0 < z \leq R$ (R sufficiently large). Writing $\tau_n = \sum_{\nu=0}^{n} (H_\chi)_{n\nu} s_\nu$ we find that $g(z) = z \sum_0^{\infty} \frac{\tau_n}{(1+z)^{n+1}}$ is regular for large $|z|$ (note that $\tau_n = O(1) \sup_{\nu \leq n} |s_\nu|$), and we have for large $|z|$

$$g(z) = z \sum_0^{\infty} \frac{1}{(1+z)^{n+1}} \sum_{\nu=0}^{n} \binom{n}{\nu} s_\nu \int_0^1 t^\nu (1-t)^{n-\nu} d\chi(t) = z \int_0^1 \sum_{\nu=0}^{\infty} \frac{s_\nu t^\nu}{(1+z)^{\nu+1}} \sum_{n=\nu}^{\infty} \binom{n}{\nu} \left(\frac{1-t}{1+z}\right)^{n-\nu} d\chi(t)$$

$$= \int_0^1 \frac{z}{t} \sum_{\nu=0}^{\infty} \frac{s_\nu}{(1+\frac{z}{t})^{\nu+1}} d\chi(t) = \int_0^1 f(\tfrac{z}{t}) d\chi(t) \qquad (f(\infty) = s_0) \quad .$$

It follows that $g(z)$ is regular for $z > 0$. Moreover, $f(x) \to s$ implies $g(x) \to s \ (x \to +0)$ because of the regularity of H_χ (we have

$$|g(x) - s| \leq \left|\int_0^\varepsilon (f(\tfrac{x}{t}) - s) d\chi(t)\right| + \left|\int_\varepsilon^1 (f(\tfrac{x}{t}) - s) d\chi(t)\right| \qquad).$$

Lemma IV. 2. Let $\chi \in V(0,1)$, $d_n = \int_0^1 t^n d\chi(t)$, $c > 0$. If $c^n d_n \to 0$, then $\chi(t) = \chi(1)$ for $\frac{1}{c} < t < 1$.

We omit the proof which is an immediate consequence of the proof of Theorem 212 in [25].

Lemma IV. 3. For numbers $z_0 \neq 0$, $0 < r < |z_0|$ let G be the region $G = \{\lambda z : |z - z_0| < r, \lambda \geq 1\}$. Let $f(z)$ be regular for $z \in G - \{z_0\}$, not regular for $z = z_0$, and assume that $\lim\limits_{z \in G, |z| \to \infty} f(z)$ exists and that $g(z) = \int_0^1 f(\frac{z}{t}) d\chi(t)$, $\chi \in V(0,1)$, (which is regular for $z \in G - \{\alpha z_0 : 0 < \alpha \leq 1\}$) has an analytic continuation onto G. Then, there is a number $\varepsilon > 0$ such that $\chi(t) = \chi(1)$ for $1-\varepsilon < t < 1$ if z_0 is a pole of $f(z)$ or if $\lim \sqrt[n]{|d_n|}$,

$$d_n = \int_0^1 t^n d\chi(t) \quad , \quad \text{exists.}$$

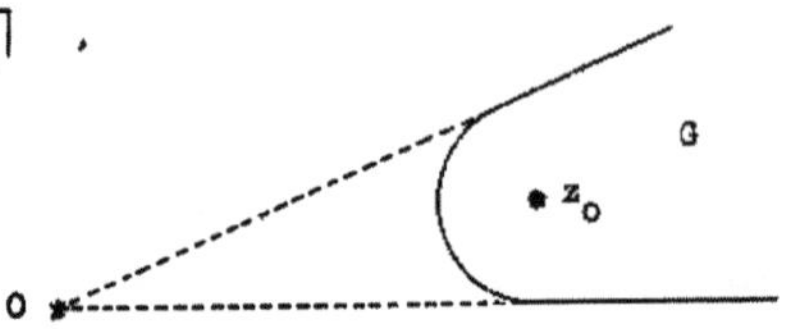

Proof. Let $f(z)$ have a representation $f(z) = \sum_{-\infty}^{\infty} c_n(z-z_0)^n$ in a neighborhood of z_0. We write $f_0(z) = f(z) - \sum_{-\infty}^{-1} c_n(z-z_0)^n = f(z) - \Phi(z)$, $z \in G$, where $\Phi(z)$ is regular for $z \neq z_0$ (its power series representation converges whenever $z \neq z_0$), and $f_0(z)$ is regular on G. It follows that $\int_0^1 f_0(\frac{z}{t})d\chi(t)$ is regular on G, and $h(z) = \int_0^1 \Phi(\frac{z}{t})d\chi(t)$ has an analytic continuation onto G. But $h(z)$ is regular for all z with the exception of the ray $\alpha z_o : \alpha \geq 0$, and it follows that $h(z)$ is regular for $|z| > |z_0| - r$.

The function $\Phi(z)$ is regular for all z (including ∞, $\Phi(\infty) = 0$) with the exception of $z = z_0$. Hence, we have a representation $\Phi(z) = \sum_1^{\infty} \frac{a_n}{z^n}$, $\limsup \sqrt[n]{|a_n|} = |z_0|$, and it follows that $h(z) = \sum_1^{\infty} \frac{a_n}{z^n} \int_0^1 t^n d\chi(t) = \sum_1^{\infty} \frac{a_n d_n}{z^n}$, where $\limsup \sqrt[n]{|a_n d_n|} \leq |z_0| - r$ because of the regularity of $h(z)$ for $|z| > |z_0| - r$.

Now let $f(z)$ have a pole at z_0, i.e.

$$\Phi(z) = \sum_{\nu=1}^{k} \frac{c_{-\nu}}{(z-z_0)^\nu} = \sum_{n=1}^{\infty} \frac{1}{z^n} \sum_{\mu=0}^{n} A_\mu^{n-\mu-1} z_0^\mu c_{-(n-\mu)} \quad , \quad c_{-k} \neq 0 \quad .$$

It follows that $a_n \sim n^{k-1} z_0^{n-k} c_{-k}$ which implies $\lim \sqrt[n]{|a_n|} = |z_0|$. This result, and also the assumption that $\lim \sqrt[n]{|d_n|}$ exists, show that $\limsup \sqrt[n]{|a_n d_n|} = \limsup \sqrt[n]{|a_n|}$ $\limsup \sqrt[n]{|d_n|}$. Consequently, we have $\limsup \sqrt[n]{|d_n|} = \frac{|z_0| - r}{|z_0|}$, and it follows from Lemma IV. 2. that $\chi(t) = \chi(1)$ for $1 - \frac{r}{|z_0|} < t < 1$. This completes the proof (and we may choose $\varepsilon = \frac{r}{|z_0|}$).

In the following two theorems we consider regular Hausdorff methods H_χ, and there is no loss of generality when we assume that $\chi(t)$ is normalized, i.e. that $\chi(t+0) = \chi(t)$ for $0 < t < 1$.

Theorem IV. 1. Let H be a regular Hausdorff method of order ρ. Assume that $\Sigma a_n = s$ (H), and that $a_n = O(c^n)$ for some $c > 0$. Moreover, let $F(z) = \Sigma a_n z^n$ have an analytic continuation onto the circle $C_\rho : \left| z - \frac{1-\rho}{2-\rho} \right| < \frac{1}{2-\rho}$ with the exception of isolated singularities. Then F is regular on C_ρ if the singularities of F are poles or if $\lim \sqrt[n]{|d_n|}$ exists.

Proof. We write $f(z) = F\left(\frac{1}{1+z}\right) = z \; \Sigma \frac{s_n}{(1+z)^{n+1}}$, $g(z) = z \, \Sigma \frac{\tau_n}{(1+z)^{n+1}}$

$\left(\tau_n = \sum_{\nu=0} (H_\chi)_{n\nu} s_\nu \right)$, $g_1(z) = g(\rho z)$, $\chi_1(t) = \chi(\rho t)$.

The function $f(z)$ is regular with the exception of isolated singularities on $|z \rho + 1| > 1$, and $g_1(z)$ is regular on that circle (note that $g(z)$ is regular on $|z+1| > 1$). It follows from the proof of Lemma IV. 1. that $g_1(z) = \int_0^1 f(\frac{z}{t}) d\, \chi_1(t)$, and it follows from the definition of $f(z)$ that $\lim_{|z| \to \infty} f(z) = s_0$.

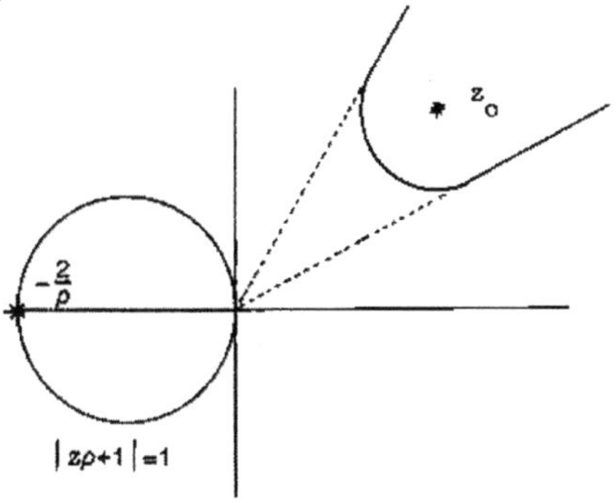

Assume that $f(z)$ has a singularity at z_0, $|\rho z_0 + 1| > 1$, and that $f(z)$ is regular for λz_0, $\lambda > 1$. Then it follows from Lemma IV. 3. that $\chi(\rho t) = \chi_1(t) = \chi_1(1) = \chi(\rho) = \chi(1)$ for $1 - \varepsilon < t \leq 1$, and this contradicts the definition of the order ρ.

We are now in a position to prove

Theorem IV. 2. ([46]) Real and regular Nörlund methods and regular Hausdorff methods are consistent.

Proof. It follows from $s_n \to s$ (N_p) by Theorem II. 5. that $F(z) = \Sigma a_n z^n$ has a positive radius of convergence, and $F(z)$ is regular on $|z| < 1$ with the exception of poles. It then follows from $s_n \to s'$ (H) by Theorem IV. 1. that $F(z)$ is regular for $\left| z - \frac{1}{2} \right| < \frac{1}{2}$ (note that C_ρ is contained in $|z| < 1$ and contains $\left| z - \frac{1}{2} \right| < \frac{1}{2}$). This implies $s_n \to s$ (A^*) and $s_n \to s'$ $(A^* H)$ (the method A^* is regular), and we have $s = s'$ by Lemma IV. 1. .

Problems.

1. Let N_p and N_q be regular and positive Nörlund methods and N_r the Nörlund method with $r_n = \sum_{\nu=0}^{n} p_{n-\nu} q_\nu$. Show that N_r is regular and $N_p \subseteqq N_r$ (with identical limits). Use this result to show that positive and regular Nörlund methods are consistent.

2. Show that $\{2^n\}$ is not limitable by a regular Hausdorff method.

2. Analytic continuation by Hausdorff methods.

In this section we will discuss the region S_A^f of Theorem III. 12. for Hausdorff methods $A = H$, and we will show that S_A^f is in many cases (but not always) the exact region of summability in the sense that no summability takes place outside of the closure $\bar{S}_A^f$ of S_A^f.

Theorem IV. 3. (R. P. Agnew [2]). Let H be a regular Hausdorff method of order ρ. Then $\sum z^n = \frac{1}{1-z}$ (H) is true for $z \in R(\rho) : |z - \frac{\rho-1}{\rho}| < \frac{1}{\rho}$, and $\sum z^n$ is not summable when z is not in the closure $\overline{R(\rho)}$ of $R(\rho)$.

Proof. (see problem 8 page 25) Let $z \in R(\rho)$.

$$\sigma_n = \sum_{\nu=0}^{n} (H)_{n\nu} z^\nu = \int_0^1 (1+t(z-1))^n d\chi(t) = \int_0^\rho (1+t(z-1))^n d\chi(t)$$

(we may assume that $\chi(t+0) = \chi(t)$ for $0 < t < 1$). If

$$\left| z - \frac{\rho - 1}{\rho} \right| < \frac{1}{\rho} \qquad \text{then} \qquad \left| z - \frac{t-1}{t} \right| < \frac{1}{t} \qquad \text{for} \qquad 0 < t \leq \rho \quad ,$$

and it follows that $|1+t(z-1)| < 1$ for $0 < t \leq \rho$, which implies

$$|\sigma_n| \leq \int_0^{\varepsilon} |1+t(z-1)|^n \, |d\chi(t)| + \int_{\varepsilon}^{\rho} |1+t(z-1)|^n \, |d\chi(t)| = o(1) \qquad (n \to \infty)$$

(observe that $|1+t(z-1)| \leq 1$ for $0 \leq t \leq \rho$ and that χ is continuous at $t = 0$) .

Before we turn to proof that Σz^n is not summable H for $z \notin \overline{R(\rho)}$ we mention that $R(\rho)$ can also be written in the form $\left| \frac{1}{z} - \frac{1-\rho}{2-\rho} \right| > \frac{1}{2-\rho}$. Now assume that Σz_0^n is summable H . It follows from Theorem IV. 1. that $F(z) = \Sigma z_0^n z^n = \frac{1}{1-z_0 z}$ is regular for $\left| z - \frac{1-\rho}{2-\rho} \right| < \frac{1}{2-\rho}$. Thus, the singularity $\frac{1}{z_0}$ cannot lie in this circle, i. e. we have $\left| \frac{1}{z_0} - \frac{1-\rho}{2-\rho} \right| \geq \frac{1}{2-\rho}$, and this implies $z_0 \in \overline{R(\rho)}$.

In connection with Theorem III.12 . we have introduced a set R . For a Hausdorff method of order ρ we write $R'(\rho)$ for R' . Starting from $R = R(\rho)$ we obtain for $R'(\rho)$ the set

$$\left| z - \frac{1-\rho}{2-\rho} \right| \leq \frac{1}{2-\rho} \quad .$$

Consider $f(z) = \Sigma a_n z^n$ with the Mittag-Leffler star S_f , then $\Sigma a_n z^n$ is summable H by Theorem III. 12. for

$$z \in S_H^f = \bigcup_{\zeta R'(\rho) \subset S_f} \zeta = \bigcap_{\zeta \notin S_f} \zeta R(\rho) \quad .$$

(A short proof shows that Σz^n is uniformly summable H on every compact subset of $R(\rho)$.)

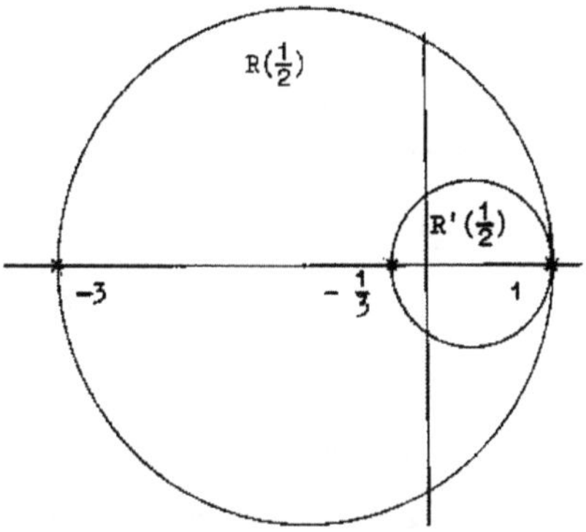

<u>Lemma IV. 4.</u> Let H be a Hausdorff method of order ρ , and let $f(z) = \Sigma a_n z^n$ be a power series with $a_n = O(c^n)$ for some $c > 0$. Then S_H^f is the exact region of summability H for $\Sigma a_n z^n$ (i. e. $\Sigma a_n z^n$ is summable H when $z \in S_H^f$, and not summable when $z \notin \overline{S}_H^f$) iff

(1) summability H of $\Sigma a_n \hat{z}^n$ implies $\hat{z} R'(\rho)_0 \subset S_f$,

where $R'(\rho)_0$ is the interior of $R'(\rho)$.

Proof. Let S_H^f be the exact region of summability. If $\Sigma a_n \hat{z}^n$ is summable, then $\hat{z} \in \overline{S}_H^f$ which implies that numbers $z_k \to \hat{z}$, $z_k \subset S_H^f$, exist, i.e. we have $z_k R'(\rho) \in S_f$, and it follows that $\hat{z} R'(\rho)_o \subset S_f$.

Assume that (1) is true, and let $\Sigma a_n \hat{z}^n$ be summable. Then $\hat{z} R'(\rho)_o \subset S_f$, and this implies $\hat{z}(1-\varepsilon) R'(\rho) \subset S_f$ for $0 < \varepsilon < 1$. Therefore, we have $(1-\varepsilon)\hat{z} \in S_H^f$ which implies $\hat{z} \in \overline{S}_H^f$.

Remark. Let H be a regular Hausdorff method of order ρ. Then we deduce from Lemma IV. 4. that S_H^f is the exact region of summability for any series $\Sigma a_n z^n$, $a_n = O(c^n)$ for some $c > 0$, iff

(2) summability H of Σc_n, $c_n = O(c^n)$ for some $c > 0$, implies that $c(z) = \Sigma c_n z^n$ is regular for $z \in R'(\rho)_o$.

(We omit the simple proof.)

We show that (2) is not always true. This follows from

Theorem IV. 4. There is a power series $\Sigma c_n z^n$ and a regular Hausdorff method H with the following properties :

(i) $\Sigma c_n z^n$ has a positive radius of convergence ,

(ii) $c(z) = \Sigma c_n z^n$ has $\left| z + \frac{1}{8} \right| = \frac{3}{8}$ as its natural boundary,

(iii) Σc_n is summable H .

Proof. Let $\chi \in V(0,1)$, $\chi(1) - \chi(0) = 1$, $\chi(+0) = \chi(0)$ be such that

$$d_{n_k} = \int_0^1 t^{n_k} \, d\chi(t) = 0 \quad \text{for natural numbers } n_k \text{ with } \frac{n_{k+1}}{n_k} \geq 1 + \lambda > 1 \quad (k=1,2,\ldots).$$

(At the end of this proof we will show that such a function χ exists.) Let $f(z) = \Sigma \frac{b_n}{z^n}$, $b_{n_k} = 3^{n_k}$, $b_n = 0$ otherwise ; this function is regular for $|z| > 3$, and it follows from Theorem III. 16. that $|z| = 3$ is a natural boundary. We write $c(\zeta) = f\left(\frac{1-\zeta}{\zeta}\right) = \Sigma c_n \zeta^n$; $c(\zeta)$ is regular for $\left|\zeta + \frac{1}{8}\right| < \frac{3}{8}$ and has $\left|\zeta + \frac{1}{8}\right| = \frac{3}{8}$ as natural boundary. It follows $\left(z = \frac{1-\zeta}{\zeta}\right)$ that

$$f(z) = \sum_0^\infty \frac{c_n}{(1+z)^n} = z \sum_0^\infty \frac{C_n}{(1+z)^{n+1}} , \qquad C_n = c_0 + \ldots + c_n$$

for $|z|$ large. Writing $\tau_n = \sum_{\nu=0}^{n} (H_\chi)_{n\nu} C_\nu$, $g(z) = z \sum_{0}^{\infty} \frac{\tau_n}{(1+z)^{n+1}}$ ($|z|$ large) we have from the proof of Lemma IV.1. the relation $g(z) = \int_0^1 f(\frac{z}{t}) d\chi(t)$, and it follows from

$$g(z) = \int_0^1 f(\tfrac{z}{t}) d\chi(t) = \Sigma \frac{b_n}{z^n} \int_0^1 t^n d\chi(t) = \Sigma \frac{b_n d_n}{z^n} \equiv 0$$

that $\tau_n = 0$, $n=0,1,\ldots$. This proves the theorem.

(It follows from the fact that the Ch. H. Müntz approximation theorem cannot (essentially) be improved - see e.g. O. Szasz [98], page 484 - that the functions t^{2^k} form not a basis in $C(0,1)$. Hence, there is a function $\phi \in V(0,1)$ such that $\int_0^1 t^{2^k} d\phi(t) = 0$, $\int_0^1 t^p d\phi(t) > 0$ for some natural number p - see e.g. S. Banach [5], page 73. We write $\chi(t) = \int_0^t t^p d\phi(t) / \int_0^1 t^p d\phi(t)$ and have $\int_0^1 t^{2^k - p} d\chi(t) = 0$ ($2^k > p$) and it follows that $n_k = 2^k - p$ is a sequence as required.)

Theorem IV.4. shows that S_H^f in general is not the exact region of summability. On the other hand, it is possible to show that S_H^f is the exact region of summability if we consider only functions of a certain class, or if we consider only a special class of methods H. A result concerning special classes of functions is :

<u>Theorem IV.5.</u> (i) Let f be a meromorphic function regular at $z=0$, and let H be a regular Hausdorff method, then S_H^f is the exact region of summability.

(ii) Let f be regular for all z with the exception of isolated singularities, and assume that f is regular for $z=0$. If H is a regular Hausdorff method and such that $\lim \sqrt[n]{|d_n|}$ exists, then S_H^f is the exact region of summability.

<u>Proof.</u> This is an immediate consequence of Theorem IV.1. and Lemma IV.4..

The Euler method E_p is a Haudorff method of order $\frac{p}{p+1}$

($\chi(t) = 0$ for $0 \leq t < \frac{p}{p+1}$, $\chi(t) = 1$ for $\frac{p}{p+1} \leq t \leq 1$).

<u>Theorem IV.6.</u> (see [54]) $S_{E_p}^f$ is the exact region of summability for the method E_p .

Proof. Let $g_\rho(\omega) = \frac{\rho\omega}{1-(1-\rho)\omega}$ $\left(\rho = \frac{p}{p+1}\right)$ and $a_n = O(c^n)$ for some $c > 0$ and define α_n by $\Sigma \alpha_n \omega^n = \Sigma a_n [g_\rho(\omega)]^n$ ($|\omega|$ small).

If $\sigma_n = \alpha_0 + \ldots + \alpha_n$, $s_n = a_0 + \ldots + a_n$, then $\sigma_n = \sum_{\nu=0}^{n} (E_p)_{n\nu} s_\nu$, as is seen from the following calculation

$$\sum_{n=0}^{\infty} \sigma_n \omega^n = \frac{1}{1-\omega} \Sigma \alpha_n \omega^n = \frac{1}{1-(1-\rho)\omega} \sum_{n=0}^{\infty} s_n [g_\rho(\omega)]^n = \sum_{n=0}^{\infty} s_n \rho^n \omega^n \sum_{m=0}^{\infty} A_m^n (1-\rho)^m \omega^m$$

$$= \sum_{k=0}^{\infty} \omega^k \sum_{n+m=k} A_m^n (1-\rho)^m \rho^n s_n = \sum_{k=0}^{\infty} \omega^k \sum_{n=0}^{k} \binom{k}{n} \rho^n (1-\rho)^{k-n} s_n .$$

Now let Σa_n be summable E_p . It follows that $\Sigma \alpha_n \omega^n$ is regular for $|\omega| < 1$. But $g_\rho(\omega)$ maps $|\omega| < 1$ onto $R'(\rho)_0$, hence $\Sigma a_n z^n$ is regular for $z \in R'(\rho)_0$ which proves the theorem because of the remark after Lemma IV. 4..

Problems.

1. Show that S_H^f is the exact region of summability if H is regular and $M_K(H)$ holds.
2. Show that S_H^f is the exact region of summability for $H = C_\alpha E_p$
3. Is S_H^f the exact region of summability for any positive regular Hausdorff method?
4. Show that $S_{HE_p}^f$ is the exact region of summability if S_H^f is the exact region of summability.

3. On convergence fields of Nörlund means

In this section we will discuss a relation between sequences limitable by a regular Nörlund mean N_p (the convergence field of N_p) and the zeros of the function $p(z) = \Sigma p_n z^n$.

First we will generalize the definition of a Nörlund mean slightly: Instead of $P_n \neq 0$ for $n \geq 0$ we require $p_0 \neq 0$ and $P_n \neq 0$ for $n \geq N$ (and σ_n is defined for $n \geq N$). Let N_p be a regular Nörlund mean with $p_0 \neq 0$, $P_n \neq 0$ for $n \geq N$, and let N_p^* be the transformation $\sigma_n = \frac{1}{P_n^*} \sum_{\nu=0}^{n} p_{n-\nu} s_\nu$, $P_n^* = P_n$ for $n \geq N$, $P_n^* = p_0$ for $0 \leq n < N$. Obviously $N_p \approx N_p^*$. The inverse of N_p^* is $s_n = \sum_{\nu=0}^{n} k_{n-\nu} P_\nu^* \sigma_\nu$, $k(z)p(z) = 1$, and an inspection of the proof of Theorem II. 7. shows that the relation $N_p^* \approx N_q^*$ holds for the two methods N_p^*, N_q^* iff the conditions of Theorem II. 7. are satisfied. Thus, Theorem II. 7. remains true for generalized Nörlund means.

Lemma IV. 5. ([83], [75]) Let N_p be regular, $q(z) = (1-\frac{z}{\alpha})p(z)$, $0 < |\alpha| < 1$, $p(\alpha) \neq 0$. Then $\{q_n\}$ generates a regular Nörlund method N_q, and $s_n \to 0$ (N_q) iff $s_n = t_n + \frac{c}{\alpha^n}$, $t_n \to 0$ (N_p), c constant.

Proof. We have $q_n = p_n - \frac{1}{\alpha} p_{n-1}$, $p_{-1} = 0$, $Q_n = P_n - \frac{1}{\alpha} P_{n-1}$, $P_{-1} = 0$.
It follows $\frac{Q_n}{P_n} = 1 - \frac{1}{\alpha}\frac{P_{n-1}}{P_n} \to 1 - \frac{1}{\alpha}$ which implies $Q_n \neq 0$ for n large. Furthermore,

$$\sum_{\nu=0}^{n} |q_\nu| \leq \sum_{\nu=0}^{n} |p_\nu| + \frac{1}{\alpha} \sum_{\nu=0}^{n} |p_\nu| = O(P_n) = O(Q_n) ,$$

and $q_n = p_n - \frac{1}{\alpha} p_{n-1} = o(P_n) = o(Q_n)$, which implies that N_q is regular.

Now let $s_n \to 0$ (N_q). If we write (formally) $s(z) = \sum s_n z^n$, then it follows for $t_n = s_n - \frac{c}{\alpha^n}$ that

$$p(z)t(z) = p(z)\left(s(z) - \frac{c}{1-\frac{z}{\alpha}} \right) = \frac{1}{1-\frac{z}{\alpha}} \left(q(z)s(z) - c\,p(z)\right)$$

where $A(z) = q(z)s(z)$ is regular for $|z| < 1$. Let $c = \frac{A(\alpha)}{p(\alpha)}$, then $A(z) - cp(z) = \sum_0^\infty c_n z^n$, $c_n = \sigma_n Q_n - c p_n$ (n large), and

$$p(z)t(z) = \sum_{n=0}^{\infty} \left(\frac{z}{\alpha}\right)^n \sum_{\nu=0}^{n} c_\nu \alpha^\nu = - \sum_{n=0}^{\infty} \left(\frac{z}{\alpha}\right)^n \sum_{\nu=n+1}^{\infty} c_\nu \alpha^\nu .$$

It follows from $\left| \frac{Q_{n+1}}{Q_n} \right| \leq (1+\varepsilon)$ (n large) that $\left|\frac{Q_\nu}{Q_n}\right| \leq (1+\varepsilon)^{\nu-n}$ $(\nu \geq n,\ n \text{ large})$ and we obtain from $|Q_\nu| = O((1+\varepsilon)^{\nu-n} P_n)$ and $c_\nu = o((1+\varepsilon)^{\nu-n} P_n)$ that

$$\sum_{\nu=0}^{n} p_{n-\nu} t_\nu = - \sum_{\nu=n+1}^{\infty} c_\nu \alpha^{\nu-n} = o(P_n) \sum_{\nu=n+1}^{\infty} ((1+\varepsilon)|\alpha|)^{\nu-n} = o(P_n)$$

which shows that $t_n \to 0 \ (N_p)$.

Finally we wish to show that $t_n \to 0 \ (N_p)$ implies $t_n \to 0 \ (N_q)$ and that $\frac{1}{\alpha^n} \to 0 \ (N_q)$.
But

$$\sum_{\nu=0}^{n} q_{n-\nu} t_\nu = \sum_{\nu=0}^{n} p_{n-\nu} t_\nu - \frac{1}{\alpha} \sum_{\nu=0}^{n-1} p_{n-1-\nu} t_\nu = o(P_n) = o(Q_n)$$

and

$$\sum_{\nu=0}^{n} q_{n-\nu} \frac{1}{\alpha^\nu} = \frac{1}{\alpha^n} \sum_{\nu=0}^{n} q_\nu \alpha^\nu = -\frac{1}{\alpha^n} \sum_{\nu=n+1}^{\infty} q_\nu \alpha^\nu = -\sum_{\nu=n+1}^{\infty} q_\nu \alpha^{\nu-n}$$

$$= -\sum_{\nu=n+1}^{\infty} o(Q_\nu)\alpha^{\nu-n} = o(Q_n) \sum_{\nu=n+1}^{\infty} ((1+\varepsilon)|\alpha|)^{\nu-n} = o(Q_n) .$$

Lemma IV. 6. ([75]) Let N_p be regular , $r(z) = \Sigma r_n z^n$, $\Sigma |r_n| < \infty$, $r(1) \neq 0$. Then, if $q(z) = r(z)p(z)$, the sequence $\{q_n\}$ generates a regular Nörlund method, and $N_p \approx N_q$ if $r(z) \neq 0$ for $|z| \leq 1$.

Proof. We have $q_n = \sum_{\nu=0}^{n} p_{n-\nu} r_\nu$, $Q_n = \sum_{\nu=0}^{n} p_{n-\nu} R_\nu$ $(R_n = r_0 + \dots + r_n)$. It follows from the regularity of N_p that $\frac{Q_n}{P_n} = \frac{1}{P_n} \sum_{\nu=0}^{n} p_{n-\nu} R_\nu \to r(1)$, and this yields

$$q_n = P_n\left(\frac{1}{P_n} \sum_{\nu=0}^{n} p_{n-\nu} r_\nu\right) = o(P_n) = o(Q_n)$$

and

$$\sum_{\nu=0}^{n} |q_\nu| \leq \sum_{\nu=0}^{n} |p_\nu| \sum_{\mu=\nu}^{n} |r_{\mu-\nu}| = O(P_n) = O(Q_n)$$

which implies that N_q is regular.

The relation $N_p \approx N_q$ follows from Theorem II. 7. , provided that $\frac{1}{r(z)} = \Sigma \rho_n z^n$ satisfies $\Sigma |\rho_n| < \infty$. This is under our assumptions on $r(z)$ a consequence of a Theorem of Wiener and Levy (see [116]).

The following theorem is an immediate consequence of Lemmas IV. 5. and IV. 6..

Theorem IV. 7. ([75]) Let N_p be regular. Assume that $r(z) = \Sigma r_n z^n$, $\Sigma |r_n| < \infty$, $r(z) \neq 0$ for $|z| \leq 1$, and that $\alpha_1, \dots, \alpha_k$ are complex numbers with $0 < |\alpha_i| < 1$, $\alpha_i \neq \alpha_j$ for $i \neq j$, $p(\alpha_i) \neq 0$. Then $q(z) = (1 - \frac{z}{\alpha_1}) \dots (1 - \frac{z}{\alpha_k})\, r(z)p(z)$ defines a regular Nörlund mean N_q , and $s_n \to 0 \ (N_q)$ iff

$$s_n = t_n + \frac{c_1}{\alpha_1^n} + \ldots + \frac{c_k}{\alpha_k^n} \quad , \quad t_n \to 0 \ (N_p) \ , \quad c_1, \ldots, c_k \quad \text{constant} \quad .$$

As an application of Theorem IV. 7. we discuss the relation between the convergence fields of the discontinuous Riesz means (R^*, n, κ) and the Cesàro means C_κ . The Riesz method (R^*, n, κ) is a Nörlund mean N_q with $q_n = (n+1)^\kappa - n^\kappa$, C_κ is a Nörlund mean N_p with $p_n = A_n^{\kappa-1}$. Here $q(z) = \sum_{n=0}^{\infty} ((n+1)^\kappa - n^\kappa) z^n = (1-z) \sum_0^\infty (n+1)^\kappa z^n = (1-z) f_\kappa(z)$, and $p(z) = (1-z)^{-\kappa}$.

If $\kappa = k = 1, 2, \ldots$, then $f_k(z) = (\frac{d}{dz} z)^k \frac{1}{(1-z)} = \frac{P_{k-1}(z)}{(1-z)^{k+1}}$ where $P_{k-1}(z)$ is a polynomial of degree $\leq k-1$ ($P_0(z) = 1$, $P_1(z) = 1+z$, $P_2(z) = 1+4z+z^2$, $P_3(z) = z^3+11z^2+11z+1$, $P_4(z) = 1+26z+66z^2+26z+1$). A short proof shows that $P_k(z) = z^k P_k(1/z)$ - which implies that $P_{2k+1}(-1) = 0$ - and that all zeros of P_{2k} and P_{2k+1} are simple, real, and negative and that k zeros are located in each of the intervals $(-1, 0)$, $(-\infty, -1)$ (see e.g. [63], [83]).

It follows from Theorem IV. 7. that $s_n \to 0 \ (R^*, n, 2k+1)$ iff

$$s_n = t_n + \frac{c_1}{\alpha_1^n} + \ldots + \frac{c_k}{\alpha_k^n} \quad , \quad t_n \to 0 \quad (C_{2k+1}) \ , \quad \text{where} \quad \alpha_1, \ldots, \alpha_k$$

are those zeros of P_{2k} which are located in $(-1, 0)$.
For the relation between (R^*, n, κ) and C_κ for all $\kappa > 0$ see [75] .

In conclusion we will show that $(R^*, n, \kappa) \approx C_\kappa$ for $0 < \kappa < 2$.

Lemma IV. 7. ([59], for the following proof see [85]) If $0 < \kappa < 2$, then $f_\kappa(z) \neq 0$ for $|z| \leq 1$ ($f_\kappa(1) = \infty$) . (The following proof will show that $\lim_{r \to 1} f_\kappa(re^{i\phi}) = f_\kappa(e^{i\phi})$ exists for $\phi \not\equiv 0 \ (2\pi)$)

Proof. We have $(1-z)^2 f_\kappa(z) = 1 + \sum_{n=1}^{\infty} ((n+1)^\kappa - 2n^\kappa + (n-1)^\kappa) z^n = 1 + \sum_{n=1}^{\infty} \Delta^2 (n-1)^\kappa z^n$. Here,

$$\Delta^m \Delta^2 (n-1)^\kappa = \Delta^{m+2} (n-1)^\kappa = \kappa(\kappa-1) \ldots (\kappa-m-1) \, \theta^{\kappa-m-2} (-1)^{m+2}$$

for some $n-1 < \theta < n+m+2$, and it follows that

$$\Delta^{m+2} (n-1)^\kappa \geq 0 \quad 1 \leq \kappa < 2 \quad , \quad \Delta^{m+2} (n-1)^\kappa \leq 0 \quad 0 < \kappa < 1 \quad .$$

Consequently, by the Hausdorff moment problem (see page 23)

$$\Delta^2 (n-1)^{\kappa} = \int_0^1 t^n dg(t) \quad , \quad g(t) \downarrow \text{ for } 0<\kappa<1 \quad , \quad g(t) \uparrow \text{ for } 1 \le \kappa < 2 \quad ,$$

and

$$(1-z)^2 f_{\kappa}(z) = 1 + z \int_0^1 \frac{dg(t)}{1-zt}$$

(and this representation shows that $f_{\kappa}(z)$ admits analytic continuation onto the domain $\{ z : \operatorname{Im} z \neq 0 \text{ if } \operatorname{Re} z \ge 1 \}$). We have for $z = re^{i\theta}$

$$1 + z \int_0^1 \frac{dg(t)}{1-zt} = 1 + r \int_0^1 \frac{\cos\theta - rt}{|1-zt|^2} dg(t) + i r \sin\theta \int_0^1 \frac{dg(t)}{|1-zt|^2} \quad ,$$

and this shows that $(1-z)^2 f_{\kappa}(z)$ has no non-real zero. Furthermore, if $z = x$ is real

$$(1-x)^2 f_{\kappa}(x) = 1 + \int_0^1 \frac{x}{1-xt} dg(t) \quad ,$$

and $\frac{x}{1-xt} \uparrow$ for $x \uparrow$, which implies that

$$(1-x)^2 f_{\kappa}(x) \uparrow \quad \text{for} \quad 1 \le \kappa < 2 \; , \quad (1-x)^2 f_{\kappa}(x) \downarrow \quad \text{for} \quad 0 < \kappa \le 1 \; .$$

Finally,

$$(1-z)^2 f_{\kappa}(z) \Big|_{z=-1} = 4 \, \zeta(-\kappa)(1-2^{1+\kappa}) > 0 \qquad (1 < \kappa < 2)$$

(ζ denotes Riemann's zeta function) , and

$$\lim_{x \to 1} (1-x)^2 f_{\kappa}(x) = 0 \qquad (0 < \kappa < 1) \; ,$$

which proves the lemma.

<u>Theorem IV.8.</u> ([92], [59]) $(R^*, n, \kappa) \approx C_{\kappa}$ for $0 < \kappa < 2$.

<u>Proof.</u> In view of Lemmas IV.6. and IV.7. we have to show that $\Sigma |r_n| < \infty$, where $\frac{q(z)}{p(z)} = (1-z)^{\kappa+1} f_{\kappa}(z) = \Sigma r_n z^n$.

We have $r_n = \sum_{\nu=0}^{n} A_{n-\nu}^{-\kappa-2} (\nu+1)^{\kappa}$, and it follows from $(\nu+1)^{\kappa} = c_1 A_{\nu}^{\kappa} + c_2 A_{\nu}^{\kappa-1} + c_3 A_{\nu}^{\kappa-2} + O(\nu^{\kappa-3})$ (see for instance, [25] , 6.10.2) that

$$r_n = c_1 A_n^{-1} + c_2 A_n^{-2} + c_3 A_n^{-3} + O(1) \sum_{\nu=0}^{n} (n+1-\nu)^{-\kappa-2} (\nu+1)^{\kappa-3} ,$$

and $\Sigma \left| r_n \right| < \infty$ because of

$$\sum_n \sum_{\nu=0}^{n} (n+1-\nu)^{-\kappa-2} (\nu+1)^{\kappa-3} = \sum_{\nu=0}^{\infty} (\nu+1)^{\kappa-3} \sum_{n=\nu}^{\infty} (n+1-\nu)^{-\kappa-2} < \infty .$$

Bibliography

The following abbreviations are used

AM	Acta Mathematica
Annals	Annals of Mathematics
Arch. M.	Archiv der Mathematik
AUH	Acta litt. ac. sci. Univ. Hungaricae (Szeged)
BAMS	Bulletin of the American Mathematical Society
CJM	Canadian Journal of Mathematics
CR	Comptes rendus de l' Académie des sciences
DMJ	Duke Mathematical Journal
JLMS	Journal of the London Mathematical Society
JM	Journal für die reine und angewandte Mathematik
MA	Mathematische Annalen
MZ	Mathematische Zeitschrift
PLMS	Proceedings of the London Mathematical Society
PMF	Prace matematyczno-fizyczne (Warszawa)
RP	Rendiconti del circolo matematico di Palermo
SM	Studia Mathematica
TAMS	Transactions of the American Mathematical Society
TMJ	Tôhoku Mathematical Journal

[1] Abel, N. H. : Untersuchungen über die Reihe ...... JM 1(1826), 311-339

[2] Agnew, R. P. : Analytic extension by Hausdorff methods. TAMS 52(1942), 217-237

[3] Ananda-Rau, K. : On Lambert's series. PLMS (2), 19(1920), 1-20

[4] Andersen, A. F. : Studier over Cesàro's summabilitetsmetode. Dissertation, Copenhagen 1921

[5] Banach, S. : Théorie des opérations linéaires. Warszawa 1932

[6] Basu, S. K. : Note on some theorems on the Hölder and Cesàro means. Bull. Calcutta Math. Soc. 40(1948), 129-134

[7] Bohr, H. : Sur la série de Dirichlet. CR 148(1909), 75-80

[8] Borel, E. : Leçons sur les séries divergentes. Paris 1901

[9] Bosanquet, L. S. : A mean value theorem. JLMS 16(1941), 146-148

[10] ———— : Note on the Bohr-Hardy theorem. JLMS 17(1942), 166-173

[11] ———— : Note on convexity theorems. JLMS 18(1943), 239-248

[12] ———— : Note on the converse of Abel's theorem. JLMS 19(1944), 161-168

[13] ———— : Note on convergence and summability factors (III). PLMS (2), 50(1949), 482-496

[14] Cesàro, E. : Sur la multiplication des séries. Bull. sc. math. (2), 14(1890), 114-120

[15] Chandrasekharan, K. and Minakshisundaram, S. : Typical means. Oxford 1952

[16] Chow, H. C. : A note on a summable series. JLMS 27(1952), 352-355

[17] Fejér, L. : Untersuchungen über Fouriersche Reihen. MA 58(1904), 51-69

[18] Frobenius, G. : Über die Leibnizsche Reihe. JM 89(1880), 262-264

[19] Gaier, D. : Der allgemeine Lückensatz für das Borel-Verfahren. MZ 88(1965), 410-417

[20] Gaier, D. und Peyerimhoff, A. : Summierbarkeitsfaktoren bei Eulerschen Reihentransformationen. MZ 58(1953), 232-242

[21] Garabedian, H. L. and Randels, W. C. : Theorems on Riesz means. DMJ 4(1938), 529-533

[22] Hardy, G. H. : Generalisation of a theorem in the theory of divergent series. PLMS (2), 6(1908), 255-264

[23] ———— : Theorems relating to the summability and convergence of slowly oscillating series. PLMS (2), 8(1910), 301-320

[24] ———— : Generalisations of a limit theorem of Mr. Mercer. PLMS(2), 6(1908), 255-264

[25] ———— : Divergent series. Oxford 1959

[26] Hardy, G. H. and Littlewood, J. E. : The relations between Borel's and Cesàro's methods of summation. PLMS (2), 11(1913), 1-16

[27] ———— : Contributions to the arithmetic theory of series. PLMS (2), 11(1913), 411-478

[28] ———— : Theorems concerning the summability of series by Borel's exponential method. RP 41(1916), 36-53

[29] ———— : On a Tauberian theorem for Lambert's series, and some fundamental theorems in the analytic theory of numbers. PLMS (2), 19(1921), 21-29

[30] ———— : A further note on the converse of Abel's theorem. PLMS (2), 25(1926), 219-236

[31] Hardy, G. H. and Riesz, M. : The general theory of Dirichlet's series. Cambridge 1915

[32] Hausdorff, F. : Summationsmethoden und Momentfolgen. MZ 9(1921), 74-109

[33] Hill, J. D. : On perfect methods of summability. DMJ 3(1937), 702-714

[34] ———— : Some properties of summability II. BAMS 50(1944), 227-230

[35] Hobson, E. W. : The theory of functions of a real variable II. Cambridge 1926

[36] Hoischen, L. : Beiträge zur Limitierungstheorie. Mitteilungen aus dem Math. Sem. Giessen 1962

[37] Hölder, O. : Grenzwerthe von Reihen an der Convergenzgrenze. MA 20(1882), 535-549

[38] Hurwitz, W. A. and Silverman, L. L. : On the consistency and equivalence of certain definitions of summability. TAMS 18(1917), 1-20

[39] Jurkat, W. : Über Rieszsche Mittel mit unstetigem Parameter. MZ 55(1951/52), 8-12

[40] ———— : Über Rieszsche Mittel und verwandte Klassen von Matrixtransformationen. MZ 57(1952/53), 353-394

[41] ———— : Ein funktionentheoretischer Beweis für O-Taubersätze bei Potenzreihen. Arch. M. 7(1956), 122-125

[42] ———— : Ein funktionentheoretischer Beweis für O-Taubersätze bei den Verfahren von Borel und Euler-Knopp. Arch. M. 7(1956), 278-283

[43] Jurkat, W. und Peyerimhoff, A. : Mittelwertsätze bei Matrix- und Integraltransformationen. MZ 55(1951/52), 92-108

[44] ———— : Mittelwertsätze und Vergleichssätze für Matrixtransformationen. MZ 56(1952), 152-178

[45] ———— : Summierbarkeitsfaktoren. MZ 58(1953), 186-203

[46] ———— : The consistency of Nörlund and Hausdorff methods. Annals 62(1955), 498-503

[47] Jurkat, W. und Peyerimhoff, A. : Über Äquivalenzprobleme und andere limitierungstheoretische Fragen bei Halbgruppen positiver Matrizen. MA 159(1965), 234-251

[48] ———————— : Über Sätze vom Bohr-Hardy'schen Typ. TMJ (2), 17 (1965), 55-71

[49] Karamata, J. : Über die Hardy-Littlewoodschen Umkehrungen des Abelschen Stetigkeitssatzes. MZ 32(1930), 319-320

[50] Knopp, K. : Grenzwerte von Reihen bei der Annäherung an die Konvergenzgrenze. Dissertation, Berlin 1907

[51] ——— : Über Lambertsche Reihen. JM 142(1913), 283-315

[52] ——— : Über das Eulersche Summierungsverfahren. MZ 15(1922), 226-253

[53] ——— : Über das Eulersche Summierungsverfahren II. MZ 18(1923), 125-156

[54] ——— : Über Polynomentwicklungen im Mittag-Lefflerschen Stern durch die Anwendung der Eulerschen Reihentransformation. AM 47(1926), 313-335

[55] ——— : Zur Theorie der Limitierungsverfahren I, II. MZ 31(1930), 97-127, 276-305

[56] ——— : Theorie und Anwendung der unendlichen Reihen. Springer 1947

[57] Knopp, K. and Lorentz, G.G. : Beiträge zur absoluten Limitierung. Arch. M. 2(1949), 10-16

[58] Kojima, T. : On generalized Toeplitz's theorems on limit and their applications. TMJ 12(1917), 291-326

[59] Kuttner, B. : On discontinuous Riesz means of type n . JLMS 37(1962), 354-364

[60] ——— : A Tauberian theorem for discontinuous Riesz menas (II). JLMS 39(1964), 643-648

[61] Landau, E. : Über die Bedeutung einiger neuen Grenzwertsätze der Herren Hardy und Axer. PMF 21(1910), 97-177

[62] ——— : Über einen Satz des Herrn Littlewood. RP 35(1913), 265-276

[63] Lawden, D. F. : The function $\sum_{n=1}^{\infty} n^r z^n$ and associated polynomials. Proc. Cambridge Philos. Soc. 47(1951), 309-314

[64] Littlewood, J. E. : The converse of Abel's theorem on power series. PLMS (2), 9(1911), 434-448

[65] Lorentz, G. G. : A contribution to the theory of divergent sequences. AM 80(1948), 167-190

[66] ——— : Direct theorems on methods of summability. CJM 1(1949), 305-319

[67] ——— : Direct theorems on methods of summability II. CJM 3(1951), 236-256

[68] ——— : Bernstein polynomials. Toronto 1953

[69] Lorentz, G. G. and Macphail, M. S. : Direct theorems on methods of summability III : Absolute summability functions. MZ 59(1953/54), 231-246

[70] Mazur, S. : Eine Anwendung der Theorie der Operationen bei der Untersuchung der Toeplitzschen Limitierungsverfahren I. SM 2(1930), 40-50

[71] Mazur, S. et Orlicz, W. : Sur les méthodes linéaires de sommation. CR 196(1933), 32-34

[72] ———— : On linear methods of summability. SM 14(1954), 129-160

[73] Melikov, H. H. : A class of summation methods for divergent series. (Russian) Sev.-Osetin. Gos. Ped. Inst. Učen. Zap. 26(1964), 19-27 and Kabardino-Balkarsk. Gos. Univ. Učen. Zap. 24(1965), 183-188

[74] Mercer, J. : On the limits of real variants. PLMS (2), 5(1907), 206-224

[75] Miesner, W. : The convergence fields of Nörlund means. PLMS (3), 15(1965), 495-507

[76] Okada, Y. : Über die Annäherung analytischer Funktionen. MZ 23(1925), 62-71

[77] Ostrowski, A. : Über eine Eigenschaft gewisser Potenzreihen mit unendlich vielen verschwindenden Koeffizienten. Sitzungsber. d. Königl. Preuss. Akademie d. Wiss. (Berlin), (1921), 557-565

[78] ———— : On representation of analytical functions by power series. JLMS 1(1926), 251-263

[79] Petersen, G. E. : Convergence and summability factors. Dissertation, University of Utah 1965

[80] Petersen, G. M. : A note on divergent series. CJM 4(1952), 445-454

[81] Peyerimhoff, A. : Konvergenz- und Summierbarkeitsfaktoren. MZ 55(1951/52), 23-54

[82] ———— : Konvergenzfaktoren beim Euler-Knoppschen Limitierungsverfahren. MZ 55(1951/52), 288-291

[83] ———— : On convergence fields of Nörlund means. Proc. Amer. Math. Soc. 7(1956), 335-347

[84] ———— : On discontinuous Riesz means. Indian J. of Math. 6(1964), 69-91

[85] ———— : On the modulus of power series of a certain type. JLMS 40(1965), 260-261

[86] ———— : On the equivalence of continuous and discontinuous Riesz means. PLMS (3), 18(1968), 349-366

[87] Ramanujan, M. S. : On summability methods of type M . JLMS 29(1954), 184-189

[88] Riesz, M. : Sur les séries de Dirichlet et les séries entières. CR 149(1909), 909-912

[89] ———— : Une méthode de sommation équivalente à la méthode des moyennes arithmétiques. CR 152(1911), 1651-1654

[90] ———— : Sur la sommation des séries de Fourier. AUH 1(1922/23), 104-113

[91] Riesz, M. : Sur un théorème de la moyenne et ses applications. AUH 1(1922/23), 114-126

[92] ——— : Sur l'équivalence de certaines méthodes de sommation. PLMS (2), 22(1924), 412-419

[93] Schmidt, R. : Über divergente Folgen und lineare Mittelbildungen. MZ 22(1925), 89-152

[94] ——— : Die Umkehrsätze des Borelschen Summierungsverfahrens. Schriften Königsberg 1(1925), 205-256

[95] Schnee, W. : Die Identität des Cesàroschen und Hölderschen Grenzwertes. MA 67(1909), 110-125

[96] Schur, I. : Über die Äquivalenz der Cesàroschen und Hölderschen Mittelwerte. MA 74(1913), 447-458

[97] ——— : Über lineare Transformationen in der Theorie der unendlichen Reihen. JM 151(1921), 79-111

[98] Szász, O. : Über die Approximation stetiger Funktionen durch lineare Aggregate von Potenzen. MA 77(1916), 482-496

[99] ——— : Verallgemeinerung eines Littlewoodschen Satzes über Potenzreihen. JLMS 3(1928), 254-262

[100] ——— : On the product of two summability methods. Ann.de la Soc. Pol. de Math. 25 (1952), 75-84

[101] Tauber, A. : Ein Satz aus der Theorie der unendlichen Reihen. Monatshefte f. Math. 8(1897) 273-277

[102] Toeplitz, O. : Über allgemeine lineare Mittelbildungen. PMF 22(1913), 113-119

[103] Ullrich, E. : Zur Korrespondenz zweier Klassen von Limitierungsverfahren. MZ 25(1926), 382-387

[104] Vijayaraghavan, T. : A Tauberian theorem. JLMS 1(1926), 113-120

[105] ——— : A theorem concerning the summability of series by Borel's method. PLMS (2), 27(1928), 316-326

[106] Widder, D. V. : The Laplace transform. Princeton 1946

[107] Wiener, N. : Tauberian theorems. Annals (2), 33(1932), 1-100

[108] ——— : The Fourier integral and certain of its applications. Cambridge 1933

[109] Wilansky, A. : A necessary and sufficient condition that a summability method be stronger than convergence. BAMS 55(1949), 914-916

[110] Zeller, K. : Allgemeine Eigenschaften von Limitierungsverfahren. MZ 53(1950/51), 463-487

[111] ——— : Abschnittskonvergenz in FK-Räumen. MZ 55(1951/52), 55-70

[112] ——— : Merkwürdigkeiten bei Matrixverfahren; Einfolgenverfahren. Arch. M. 4(1953),1-

[113] Zeller, K. : Abschnittsabschätzungen bei Matrixtransformationen. MZ 80(1962/63), 355-357

[114] Zygmund, A. : O teorji srednich arythmetycznych. Mathesis Polska 1(1926), 75-85, 119-129

[115] ——— : On a theorem of Ostrowski. JLMS 6(1931), 162-163

[116] ——— : Trigonometric series I. Cambridge 1959

Additional literature.

Bromwich, T. J. I. A. : An introduction to the theory of infinite series. Cambridge 1926

Cooke, R. G. : Infinite matrices and sequence spaces. London 1950

Dienes, P. : The Taylor series. Oxford 1931

Ford, W. B. : Studies on divergent series and summability. University of Michigan Studies scientific series. Vol. 2 , New York 1916

Karamata, J. : Sur les théorèms inverses des procédés de sommabilité. Actual. sci.industr. No. 450, Paris 1937

Knopp, K. : Neuere Untersuchungen in der Theorie der divergenten Reihen. Jahresber. Deutsch. Math. Vereinigung 22(1923), 43-67

Kogbetliantz, E. : Sommation des séries et intégrales divergentes par les moyennes arithmétiques et typiques. Mémorial des sc. math. No. 51, Paris 1931

Landau, E. : Darstellung und Begründung einiger neuerer Ergebnisse der Funktionentheorie. Berlin 1929

Moore, C. N. : Summable series and convergence factors. Amer. Math. Soc. Colloquium Publ. No. 22, New York 1938

Petersen, G. M. : Regular matrix transformations. London 1966

Petersen, G. M. : Regular matrices and bounded sequences. Jahresber. Deutsch. Math. Vereinigung 69(1967)

Pitt, H. R. : Tauberian theorems. Bombay 1958

Smail, L. L. : History and synopsis of the theory of summable infinite processes. Oregon 1925

Szász, O. : Introduction to the theory of divergent series. New York 1952

Zeller, K. : Theorie der Limitierungsverfahren. Springer 1958

Index

MIX
Papier aus verantwortungsvollen Quellen
Paper from responsible sources
FSC® C105338

If you have any concerns about our products,
you can contact us on
ProductSafety@springernature.com

In case Publisher is established outside the EU,
the EU authorized representative is:
Springer Nature Customer Service Center GmbH
Europaplatz 3, 69115 Heidelberg, Germany

Printed by Libri Plureos GmbH
in Hamburg, Germany